Ten Essays on Fizz Buzz

Meditations on Python, mathematics, science, engineering, and design

Joel Grus

Ten Essays on Fizz Buzz

Meditations on Python, mathematics, science, engineering, and design

Joel Grus

ISBN 978-0-9824818-2-0

Contents

Introduction

Fizz Buzz is the following (simple) problem:

> Print the numbers from 1 to 100, except that if the number is divisible by 3, instead print "fizz"; if the number is divisible by 5, instead print "buzz"; and if the number is divisible by 15, instead print "fizzbuzz".

It originated as a children's game, but has since taken on a new life as a lowest-common-denominator litmus test for assessing computer programmers.

If you are an experienced programmer, it is an extremely easy problem to solve. Because of this, it has taken on a third life as the prototypical *bad* interview problem. Everyone knows that it's the question you ask people to make sure that they're not completely incompetent as programmers. Accordingly, if your interviewer asks you to solve it, he's suggesting he thinks it possible that you're completely incompetent as a programmer. You would not be wrong to feel insulted!

My association with this problem began in 2016, when I wrote a blog post called Fizz Buzz in Tensorflow, the (possibly fictional) story of one such insulted programmer who decided to show up his interviewer by approaching Fizz Buzz as a deep learning problem. This post went modestly viral, and ever since then I have been seen as a thought leader in the Fizz Buzz space.

Accordingly, over the years I have come up with and/or collected various other stupid and/or clever ways of solving Fizz Buzz. I have not blogged about them, as I am not the sort of person who beats a joke to death, but occasionally I will tweet about them, and recently in response someone suggested that I write a book on "100 Ways of Writing Fizz Buzz in Python."

Now, I could probably come up with 100 ways of solving Fizz Buzz, but most of them would not be very interesting. Luckily for you, I was able to come up with 10 that *are* interesting in various ways, each of which turned out to be a good launching-off point for (sometimes meandering) discussions of various aspects of coding, Python,

Fizz Buzz, mathematics, software design, technical interviewing, and various other topics. Hence "Ten Essays on Fizz Buzz". [1]

In many ways this is a strange book. Its goal is not to teach you a specific field or a specific technology. I hope you will learn a lot from reading it, but it's not really a book that you'd read in order to learn anything in particular. Most technical books are about specific technical topics; this one sort of isn't.

Nonetheless, it is a technical book. Each essay contains code that implements a different solution of Fizz Buzz. Each essay uses code to illustrate its ideas. Each essay represents my current best thinking about how to solve problems using code. If you have a coding job, you should feel no reluctance to expense this book to your employer.

As I primarily code in Python, all of these solutions will be in Python. Some of them will use features that were only introduced in Python 3.6, and some of them will demonstrate concepts and techniques that are mostly specific to Python. On some level this is deeply a *Python* book. However, my goal was to write a book that would be interesting and enlightening even if you are not a Python programmer. I'll leave it to you to judge how successful I was.

If I have explored more ways of solving Fizz Buzz than others, it is because I stood on the shoulders of giants. The "if / elif / elif / else" solution is the canonical one; the "100 print statements" solution is obvious; the "random guessing" solution I learned about from a Stack Overflow question; the remaining solutions I'm pretty sure I came up with myself, although for many of them it is a near certainty that other people have independently come up with similar (if not basically identical) solutions. That's how programming works.

The solutions from this book are all available at github.com/joelgrus/fizzbuzz, in case you don't feel like typing them into the computer yourself.

---

[1] After I started writing this book I discovered there is a blog post "Twenty Ways to Fizz Buzz", very few of which overlap with these.

1. 100 Print Statements

Fizz Buzz originated as a game for children. The idea was that the children would sit in a circle and go around in sequence calling out the next number; substituting "fizz" or "buzz" or "fizzbuzz" as appropriate; punishing mistakes according to house rules with varying levels of cruelty.

Well, if children can come up with the correct outputs without using a computer, then so can we. This suggests what is possibly the least imaginative solution – figure out the correct outputs by hand and explicitly print each one:

```
print('1')
print('2')
print('fizz')
print('4')
print('buzz')
print('fizz')
print('7')
print('8')
print('fizz')
print('buzz')
print('11')
print('fizz')
print('13')
print('14')
print('fizzbuzz')
print('16')
print('17')
print('fizz')
print('19')
print('buzz')
print('fizz')
```

```python
print('22')
print('23')
print('fizz')
print('buzz')
print('26')
print('fizz')
print('28')
print('29')
print('fizzbuzz')
print('31')
print('32')
print('fizz')
print('34')
print('buzz')
print('fizz')
print('37')
print('38')
print('fizz')
print('buzz')
print('41')
print('fizz')
print('43')
print('44')
print('fizzbuzz')
print('46')
print('47')
print('fizz')
print('49')
print('buzz')
print('fizz')
print('52')
print('53')
print('fizz')
print('buzz')
print('56')
print('fizz')
```

```
print('58')
print('59')
print('fizzbuzz')
print('61')
print('62')
print('fizz')
print('64')
print('buzz')
print('fizz')
print('67')
print('68')
print('fizz')
print('buzz')
print('71')
print('fizz')
print('73')
print('74')
print('fizzbuzz')
print('76')
print('77')
print('fizz')
print('79')
print('buzz')
print('fizz')
print('82')
print('83')
print('fizz')
print('buzz')
print('86')
print('fizz')
print('88')
print('89')
print('fizzbuzz')
print('91')
print('92')
print('fizz')
```

```
print('94')
print('buzz')
print('fizz')
print('97')
print('98')
print('fizz')
print('buzz')
```

Technically this is a solution. It solves the stated problem of printing the numbers 1 to 100 except in specific circumstances substituting "fizz" or "buzz" or "fizzbuzz".

Yet this is a solution that's not particularly satisfying. For example, we would not expect an interviewer to be impressed by it.[1]

In this first chapter we'll discuss why that's the case.

Algorithm and Abstraction

One thing the interviewer is (presumably) looking for is your ability to think algorithmically; that is, to come up with an efficient process for solving the problem by turning it into code.

But this isn't really a process for solving the problem. It's only a process for printing out a precomputed solution. It relies on already having a solution to the problem.

Nor is it particularly efficient. There is a sense in which this is about as inefficient as possible, since each of the 100 outputs is generated by its own code – nothing is shared. (There are many other senses in which there are far more inefficient solutions, and we will see some such solutions throughout the book.)

Here is a slight improvement:

---

[1]Although I would be pretty impressed if you used this solution in an interview.

```
FIZZ_BUZZ = [
    '1', '2', 'fizz', '4', 'buzz', 'fizz', '7', '8', 'fizz',
    'buzz', '11', 'fizz', '13', '14', 'fizzbuzz', '16', '17',
    'fizz', '19', 'buzz', 'fizz', '22', '23', 'fizz', 'buzz',
    '26', 'fizz', '28', '29', 'fizzbuzz', '31', '32', 'fizz',
    '34', 'buzz', 'fizz', '37', '38', 'fizz', 'buzz', '41',
    'fizz', '43', '44', 'fizzbuzz', '46', '47', 'fizz', '49',
    'buzz', 'fizz', '52', '53', 'fizz', 'buzz', '56', 'fizz',
    '58', '59', 'fizzbuzz', '61', '62', 'fizz', '64', 'buzz',
    'fizz', '67', '68', 'fizz', 'buzz', '71', 'fizz', '73',
    '74', 'fizzbuzz', '76', '77', 'fizz', '79', 'buzz',
    'fizz', '82', '83', 'fizz', 'buzz', '86', 'fizz', '88',
    '89', 'fizzbuzz', '91', '92', 'fizz', '94', 'buzz',
    'fizz', '97', '98', 'fizz', 'buzz'
]

for fizz_buzz in FIZZ_BUZZ:
    print(fizz_buzz)
```

The values are still manually specified, but at least we're using a for loop and a single call to print instead of 100 calls to print. And while we still didn't use algorithmic thinking to produce the output values, we do have the values in a list to use and re-use however we see fit.

However, this is still not a good solution, for a couple of reasons. For one thing, all of the Fizz Buzz logic here was performed by a human (me). We'd like the computer to handle that logic. What's the point of writing software if we don't let the computer do the parts it's good at?

The second reason this is not a good solution is that it's not at all *extensible*. If we suddenly needed the corresponding outputs for the numbers 101 to 200 we couldn't re-use any part of this (other than the print statement), and we'd have to do that work starting from nothing.

Nonetheless, this is an important improvement, for a couple of reasons.

Reusability and Changeability

One nice thing about solving problems with software is that (compared with many other kinds of solutions) it's relatively easy to change your solution when requirements change. For example, imagine that your interviewer decides she'd rather have the words printed in ALL CAPS.

With the "100 Print Statements" solution, you would have to go through and make that change line-by-line. Whereas with this second solution you could just make a small change:

```python
for fizz_buzz in FIZZ_BUZZ:
    print(fizz_buzz.upper())
```

Or if you wanted the results to start at 100 and count down to 1:

```python
for fizz_buzz in reversed(FIZZ_BUZZ):
    print(fizz_buzz)
```

Or if you needed to print the words without vowels ("Fzz Bzz"):

```python
import re

for fizz_buzz in FIZZ_BUZZ:
    print(re.sub("[aeiouAEIOU]", "", fizz_buzz))
```

Or if you wanted the output in Spanish:

```
for fizz_buzz in FIZZ_BUZZ:
    print(fizz_buzz
            .replace("fizz", "efervescencia")
            .replace("buzz", "zumbido"))
```

Or if you wanted to write the results out to a file:

```
with open('fizzbuzz.txt', 'w') as f:
    for fizz_buzz in FIZZ_BUZZ:
        f.write(f"{fizz_buzz}\n")
```

However, there's a limit to the sorts of modifications that this solution admits. Imagine that the interviewer decided that "buzz" should now replace multiples of 7, and "fizzbuzz" multiples of 21. There's no obvious way to modify our solution to accomplish this; probably we'd have to create an entirely new list of values:

```
FIZZ_BUZZ_7 = [1, 2, 'fizz', 4, 5, 'fizz', 'buzz', 8, ...]
```

And, as mentioned previously, if we wanted the outputs for the numbers 101 to 200, there's no way for us to reuse this work (other than the print statement).

Throughout the book we'll see various other solutions that make some of these changes much simpler. (Some of our solutions won't make these changes any simpler but will be interesting for other reasons.)

Testability and Fizz Buzz

Another important reason why the "100 print statements" solution is not great is that it's very hard to test. Is it correct? Did I make a mistake and accidentally write "fizz" when I meant to write "buzz"? The only way to know is to go through line by line and check each answer.

As we explore various ways of solving this problem, we'll want to check that our solutions are correct. After you finish this book and go on to solving other problems, you'll also want to check that your solutions are correct.

When developing software we prefer to use *automated tests*; that is, we write *test cases* that we expect to "pass" if our code is correct and that we hope will "fail" if our code is not correct.

One simple way of doing this in Python is using `assert` statements, which will raise an exception (`AssertionError`) if the condition they're asserting is false:

```python
# these assertions all pass
assert True
assert 2 > 1
assert "iz" in "FizzBuzz"

# these assertions all fail
assert False
assert 1 > 2
assert "I" in "TEAM"
```

Typically it's difficult to create test cases that are 100% comprehensive, but here we only have 100 input-output pairs to check, so we can write a test that covers every possible input / output.

We'll use the `FIZZ_BUZZ` list from this chapter as the source of truth. This means you should convince yourself that it's entirely correct.

The solutions in this book will take two general forms. Some solutions will generate a list of the 100 Fizz Buzz outputs. We'll write a function to check that such a list is correct. It first checks that the list actually has 100 elements. After that it checks that each element of the provided list is the same as the corresponding element of `FIZZ_BUZZ`.

We do that by generating a list containing all the incorrect outputs and then asserting that the list is empty. The reason we do it this way is so that when the test fails, it fails with an explicit list of all the outputs that were wrong:

```
# We need this to type-annotate lists.
from typing import List

def check_output(output: List[str]) -> None:
    assert len(output) == 100, "output should have length 100"

    # Collect all the errors in a list
    # The i+1 reflects that output[0] is the output for 1,
    # output[1] is the output for 2, and so on
    errors = [
        f"({i+1}) predicted: {output[i]}, actual: {FIZZ_BUZZ[i]}"
        for i in range(100)
        if output[i] != FIZZ_BUZZ[i]
    ]

    # And assert that the list of errors is empty
    assert not errors, f"{errors}"
```

Other solutions will result in *functions* that take in a number n and return the correct Fizz Buzz output for that n. We'll test such a function by generating the list of 100 outputs and then using our previous test. (This is another example of us re-using previous work.)

```
# We need this to type-annotate functions.
from typing import Callable

def check_function(fizz_buzz: Callable[[int], str]) -> None:
    """
    The type annotation says that `fizz_buzz` needs to be
    a function that takes a single argument (which is an `int`)
    and returns a `str`.
    """
    output = [fizz_buzz(i) for i in range(1, 101)]
    check_output(output)
```

As we explore more interesting solutions, we'll use these functions to check that they're correct, so make sure you understand them.

Testability Beyond Fizz Buzz

Throughout this book we'll apply this idea of testability to more than our various Fizz Buzz solutions. We'll also apply it to our intermediate steps and even to our digressions.

For example, imagine that one of our solutions involves checking that two words are anagrams. (None of our solutions involves this, but bear with me.)

We dutifully craft a solution of this subproblem:

```python
def anagrams(s1: str, s2: str) -> bool:
    return sorted(s1) == sorted(s2)
```

How do we know it works? By writing test cases for it:

```python
assert anagrams("dale", "lead")
assert anagrams("time", "mite")
assert not anagrams("made", "deem")
assert not anagrams("time", "miter")
```

This is not a particularly comprehensive set of tests, but it involves a couple of positive examples and a couple of negative examples. At the very least, these tests would not all pass if we'd made a stupid mistake. If they all pass, our mistake would have to be somewhat subtle.

Generating the `print` Statements

Imagine that you're writing a technical book, and that one of the chapters involves solving Fizz Buzz with 100 `print` statements, and that you don't feel like writing them out by hand.

Why not use Python to generate the `print` statements?

```python
def make_print_statement(fizz_buzz: str) -> str:
    return f"print('{fizz_buzz}')"

assert make_print_statement("10") == "print('10')"
assert make_print_statement("fizz") == "print('fizz')"
assert make_print_statement("buzz") == "print('buzz')"
assert make_print_statement("fizzbuzz") == "print('fizzbuzz')"
```

After which it's easy to print them out and then copy and paste them into your manuscript:

```python
for fizz_buzz in FIZZ_BUZZ:
    print(make_print_statement(fizz_buzz))
```

2. if / elif / elif / else

The prototypical correct solution looks something like this:

```python
def fizz_buzz(n: int) -> str:
    if n % 15 == 0:
        return 'fizzbuzz'
    elif n % 5 == 0:
        return 'buzz'
    elif n % 3 == 0:
        return 'fizz'
    else:
        return str(n)
```

Here % is the *modulus* operator, which returns the remainder from division. That is, n % 15 is the remainder you get if you divide n by 15. It equals zero precisely when n is divisible by 15 (in which case there's no remainder). Thus, n % 15 == 0 is the standard way to check whether n is divisible by 15, n % 5 == 0 the standard way to check whether n is divisible by 5, and so on.

If you were asked to solve Fizz Buzz as part of a coding interview, this is most likely the answer your interviewer would expect. If you are an experienced programmer, there is probably nothing surprising going on here. Nonetheless, there are a few aspects of this solution worth thinking about.

Mind the Order

One way to fail to solve the problem is to naively check conditions in the order they're listed in the problem description; that is, to first check "divisible by 3" then "divisible by 5" and finally divisible by 15:

```
def incorrect_fizz_buzz(n: int) -> str:
    if n % 3 == 0:
        return 'fizz'
    elif n % 5 == 0:
        return 'buzz'
    elif n % 15 == 0:
        return 'fizzbuzz'
    else:
        return str(n)
```

Any number that's divisible by 15 will trigger the first if statement and never make it to the third:

```
assert incorrect_fizz_buzz(30) == 'fizz'
```

There's a sense in which this order-dependence makes this solution fragile. At the very least it obscures the logic behind what's going on. Here's a version that doesn't depend on the order of the if statements:

```
def order_free_fizz_buzz(n: int) -> str:
    if n % 3 != 0 and n % 5 != 0:
        return str(n)
    if n % 3 == 0 and n % 15 != 0:
        return 'fizz'
    if n % 5 == 0 and n % 15 != 0:
        return 'buzz'
    if n % 15 == 0:
        return 'fizzbuzz'

    # I think we covered every case
    raise RuntimeError("impossible to get here (I hope)")
```

However, there are downsides to this version too, one of which is that it's not obvious that you've covered every case. Here it's pretty easy to mentally check that every possible input is covered by exactly one of the conditions; in a more complicated problem it might not be.

Divisibility and Modulus

This solution relies heavily on the % operator, as described above:

```python
for i in range(1, 10):
    print(f"The remainder when dividing {i} by 3 is {i % 3}.")

# The remainder when dividing 1 by 3 is 1.
# The remainder when dividing 2 by 3 is 2.
# The remainder when dividing 3 by 3 is 0.
# The remainder when dividing 4 by 3 is 1.
# The remainder when dividing 5 by 3 is 2.
# The remainder when dividing 6 by 3 is 0.
# The remainder when dividing 7 by 3 is 1.
# The remainder when dividing 8 by 3 is 2.
# The remainder when dividing 9 by 3 is 0.
```

One criticism of Fizz Buzz as a screening problem is that it's primarily a test of one's familiarity with modular arithmetic, and that a hypothetical experienced programmer who was somehow ignorant of the modulus operator would be unable to solve it (through no fault of his own).

Leaving aside the question of whether it's a good screening problem (I suspect it's not), I find this particular criticism unsatisfying, as there are plenty of other ways one could test for divisibility.

For instance, if a (positive) number is divisible by 3, then if you keep subtracting 3 from it until you get to a number less than 3, you'll end up at zero. If it's not divisible by 3, you'll end up at 1 or 2. So one way to check for divisibility is simply to carry out this process:

```
def is_divisible_subtract(n: int, d: int) -> bool:
    assert d > 0 and n >= 0

    # Keep subtracting until n < d.
    while n >= d:
        n -= d

    # If we reached zero, n was divisible by d.
    return n == 0
```

How can we be confident this works? Let's test it by checking a bunch of pairs that should be divisible

```
divisible_pairs = [(0, 3), (3, 3), (6, 3), (9, 3),
                   (0, 5), (5, 5), (10, 5), (100, 5),
                   (30, 15), (150, 15)]

for n, d in divisible_pairs:
    assert is_divisible_subtract(n, d)
```

and another set of pairs that should not be:

```
not_divisible_pairs = [(1, 3), (5, 3), (10, 3),
                       (1, 5), (2, 5), (3, 5), (4, 5),
                       (14, 15), (16, 15)]
for n, d in not_divisible_pairs:
    assert not is_divisible_subtract(n, d)
```

That's a good set of test cases, but we can do better using a library called hypothesis to do "generative testing". Rather than construct specific test cases, we specify a hypothesis that our code should satisfy (here, that is_divisible_subtract always gives the same result as checking that n % d == 0) and let the computer try to come up with counterexamples that falsify it.

You'll need to install it:

```
python -m pip install hypothesis
```

Hypothesis works by using "strategies" to generate test cases. For example, here's a test of the hypothesis that all numbers are even:

```python
from hypothesis import given
import hypothesis.strategies as st

# The hypothesis should be true for any integer n.
@given(n=st.integers())
def test_every_number_is_even(n: int):
    assert n % 2 == 0

# Normally you'd probably run your tests with pytest or similar.
# Here we'll just run it directly.
test_every_number_is_even()
```

Hypothesis quickly comes up with a falsifying example:

```
Falsifying example: test_every_number_is_even(
    n=1,
)
```

We could also test that every string is less than 20 characters long:

```python
# The hypothesis should be true for any text.
@given(s=st.text())
def test_strings_are_not_long(s: str):
    assert len(s) < 20

test_strings_are_not_long()
```

It takes a little longer, but again it finds a counterexample:

```
Falsifying example: test_strings_are_not_long(
    n='00000000000000000000',
)
```

What happens if we test something that's true?

```
@given(n=st.integers())
def test_twice_n_is_always_even(n: int):
    print(f"n: {n}")   # This will show you all the tests
    twice_n = 2 * n
    assert twice_n % 2 == 0


test_twice_n_is_always_even()
```

In this case the test passes quietly, and there is no output except from our print statement:

```
n: 0
n: 0
n: 4970
n: 0
n: -4591
n: 0
n: -29105
n: 0
n: 6200587310049051356
n: 0
n: -7063
n: -7063
n: -19645
...and a bunch more, 100 in all
```

We can use the same method to test our divisibility functions. We'll add the constraints that d needs to be positive and that n needs to be non-negative.

```
@given(n=st.integers(min_value=0), d=st.integers(min_value=1))
def test_is_divisible_subtract(n: int, d: int):
    print(f"n: {n}, d: {d}")
    assert is_divisible_subtract(n, d) == (n % d == 0)


test_is_divisible_subtract()
```

For me this test doesn't fail, but it seems to get stuck at n = 2926603197221931195, d = 35. This makes sense, since for those inputs it would take *83 quadrillion* subtractions to reach zero. If your computer could do a billion subtractions a second, that would take about 3 years. How can we deal more intelligently with such inputs?

"We're Adders, We Need Logs to Multiply"

Thinking about the logic behind our "subtract" test suggests that it still works if we subtract any multiple of d. Things would go faster if instead of just subtracting d we subtracted the largest multiple of d that we can. That is, we'd like to find the largest k such that d * k <= n and then subtract d * k.

Of course, if we knew that k, we'd know whether n was divisible by d, simply by checking whether d * k == n.

In lieu of that, we can find the largest power of d that's less than or equal to n. We'll start with d and keep multiplying by d until we get something that's larger than n:

```
def largest_power(d: int, n: int) -> int:
    """
    Finds the largest d ** k with d ** k <= n.
    """
    assert 1 < d <= n, "this only works if 1 < d <= n"

    power = d

    # Keep multiplying by d until we get something too big.
    while power * d <= n:
```

```
    power *= d

  return power
```

And, as always, we'll write a few `assert`-tests:

```
assert largest_power(3, 8) == 3
assert largest_power(3, 9) == 9
assert largest_power(3, 26) == 9
assert largest_power(3, 27) == 27
```

Of course, if you are familiar with `math.log`, you may notice a simpler way. We want to find the k that satisfies

```
d ** k <= n < d ** (k + 1)
```

If we take the log of everything that's

```
math.log(d ** k) <= math.log(n) < math.log(d ** (k + 1))
```

Then we can apply the identity: log(d ** x) == x * log(d):

```
k * math.log(d) <= math.log(n) < (k + 1) * math.log(d)
```

And divide both sides by log(d):

```
k <= math.log(n) / math.log(d) < k + 1
```

In other words, we can find the desired k by doing:

```
def largest_power(d: int, n: int) -> int:
    assert 1 < d <= n

    # int() truncates toward 0
    k = int(math.log(n) / math.log(d))
    return d ** k
```

Which we can check as usual:

```
@given(n=st.integers(min_value=1), d=st.integers(min_value=2))
def test_largest_power(n: int, d: int):
    if d <= n:
        dk = largest_power(d, n)
        assert dk <= n < dk * d

test_largest_power()
```

We can use this to create an alternative version of `is_divisible_subtract` that's fast:

```
def is_divisible_subtract_fast(n: int, d: int) -> bool:
    # all numbers are divisible by 1
    if d == 1:
        return True

    while n >= d:
        # subtract the largest power of d that's <= n
        n -= largest_power(d, n)

    # now 0 <= n < d
    return n == 0
```

And now our generative tests should pass in a reasonable amount of time:

```
@given(n=st.integers(min_value=0), d=st.integers(min_value=1))
def test_is_divisible_subtract_fast(n: int, d: int):
    assert is_divisible_subtract_fast(n, d) == (n % d == 0)


test_is_divisible_subtract_fast()
```

Some Other Checks

Another way to check that n is divisible by d is to perform the (float) division and check that the result is an integer. For example, 5 is not divisible by 2, and 5/2 is 2.5. Whereas 6 is divisible by 2, and 6/2 is 3.

```
def is_divisible_try_dividing(n: int, d: int) -> bool:
    quotient = n / d
    return quotient == int(quotient)
```

Although this works for all of our Fizz Buzz inputs, it fails for large inputs:

```
# n is pretty clearly not divisible by 2
n = 36028797018964073
assert n / 2 == int(n / 2)      # !!
assert (n + 1) / 2 == n / 2     # !!
```

Here the issue is that Python (like most computer languages) can only represent floating-point numbers in a lossy way. This means that n / 2 is only an approximation when n is large, as you can see from the second assert.

A different way is to "integer divide" (which truncates) and then multiply back and check if you get the original number.

Using the previous examples, 5 // 2 is 2, so (5 // 2) * 2 is 4 != 5. Whereas 6 // 2 is 3, so (6 // 2) * 2 is 6 == 6.

```
def is_divisible_divide_multiply(n: int, d: int) -> bool:
    return n == d * (n // d)
```

This version doesn't use floats and so it works for large numbers.

Some Well-Known Tricks

In the particular case of Fizz Buzz we really only care about whether a number is divisible by 3, by 5, or by 15. And a number is divisible by 15 precisely when it's divisible by both 3 and 5. So really we only need to be able to check whether a number is divisible by 3 and whether it's divisible by 5. And it turns out that there are well-known tricks for both.

A number is divisible by 5 if and only if its last digit is either 5 or 0. So let's write a function to get the last digit of a number and then use that:

```
def last_digit(n: int) -> int:
    # Convert to string, take last character, convert back to int
    return int(str(n)[-1])

def is_divisible_by_5(n: int) -> bool:
    # A number is divisible by 5 iff its last digit is 0 or 5
    return last_digit(n) in [0, 5]
```

And it turns out that a number is divisible by 3 if and only if the sum of its digits is divisible by 3. We'll show that this is true for 2-digit numbers (the general proof is similar).

Say we have the two-digit number n = 10a + b with digits [a, b].

Then we can write

```
n = 10a + b = (a + 9a) + b
            = (a + b) + 9a
```

No matter what a is, 9a is divisible by 3. That means that if a + b is divisible by 3, then n is the sum of two things that are divisible by 3, and so is itself divisible by 3. And if n is divisible by 3, then a + b is the difference of two things that are divisible by 3 and in that case it must be divisible by 3.

(This same proof also shows that n is divisible by 9 if and only if the sum of its digits is divisible by 9, although that piece of information doesn't really help with Fizz Buzz.)

So then we'd like to do something like:

```
def sum_of_digits(n: int) -> int:
    return sum(int(c)
                for c in str(n)
                if '0' <= c <= '9')

def is_divisible_by_3_wrong(n: int) -> bool:
    return is_divisible_by_3_wrong(sum_of_digits(n))
```

Unfortunately, this will run forever. For instance, if you called it on n = 18, you would compute:

```
is_divisible_by_3_wrong(18)   # sum of digits is 9
is_divisible_by_3_wrong(9)    # sum of digits is 9
is_divisible_by_3_wrong(9)    # sum of digits is 9
is_divisible_by_3_wrong(9)    # sum of digits is 9
# ... forever
```

So we need to add some sort of *base case* in order for the recursion to terminate. What should the base case be? Well, I insist that the following are true:

1. If a number has more than 2 digits, its sum_of_digits has fewer digits than it does.

2. If a number has 2 digits, its `sum_of_digits` is at most 18.
3. If a number is 18 or less, its `sum_of_digits` is at most 9.

The first should be easy to convince yourself of, since the sum of digits of a n-digit number is less than 10 * n (each of the n digits is at most 9), and 10 * n doesn't have 3 digits until n = 10, and doesn't have 4 digits until n = 100, and so on.

The second and third claims can be proved by *exhaustion*; that is, by just checking every possibility.

Anyway, the upshot of all this is that if we start with any positive number and keep applying `sum_of_digits` we will eventually end up with a number that's less than 10. We know which numbers less than 10 are divisible by 3, which gives us a base case to use.

For instance, this rule tells us that 999 is divisible by 3 if and only if 27 (its sum of digits) is divisible by 3, which is true if and only if 9 (its sum of digits) is divisible by 3, which it is.

```python
def is_divisible_by_3(n: int) -> bool:
    while n >= 10:
        n = sum_of_digits(n)

    # Now n is a single-digit number,
    # and we know what to do for those:
    return n in [0, 3, 6, 9]
```

Now we can write our modulus-free version:

```python
def fizz_buzz(n: int) -> str:
    by3 = is_divisible_by_3(n)
    by5 = is_divisible_by_5(n)

    if by3 and by5:
        return "fizzbuzz"
    elif by5:
        return "buzz"
    elif by3:
        return "fizz"
    else:
        return str(n)
```

Quite plainly this is not a test of whether you know the modulus operator. (Honestly, who knows what it's a test of.)

Eliminating `elif`

Some people find the four-branched `if` statement distasteful. There are various ways around it, one simple one is to replace it with a `dict` lookup:

```python
def fizz_buzz(n: int) -> str:
    by3 = is_divisible_by_3(n)
    by5 = is_divisible_by_5(n)

    return {
        (True, True): 'fizzbuzz',
        (True, False): 'fizz',
        (False, True): 'buzz',
        (False, False): str(n)
    }[by3, by5]
```

although I prefer the `if` statements myself.

A way I like better is to hide the `if`s inside a call to `list.index`, which returns the index of the first occurrence of a value:

```
def fizz_buzz(n: int) -> str:
    idx = [n % 15, n % 5, n % 3, 0].index(0)
    return ["fizzbuzz", "buzz", "fizz", str(n)][idx]
```

For example, if n were 10, that first list would be

```
[10 % 15, 10 % 5, 10 % 3, 0]
```

which is

```
[10, 0, 1, 0]
```

and so the first 0 is at index 1, corresponding to "buzz".

As a bonus, this solution is likely to confuse your interviewer.

3. The Cycle of 15

Here's a slightly less obvious variant:

```python
CYCLE_OF_15 = ['fizzbuzz', None, None, 'fizz',
               None, 'buzz', 'fizz', None,
               None, 'fizz', 'buzz', None,
               'fizz', None, None]

def fizz_buzz(n: int) -> str:
    return CYCLE_OF_15[n % 15] or str(n)
```

We certainly got rid of those pesky `if` statements. But what exactly is this code doing?

Equivalence Classes

If you ponder the Fizz Buzz problem long enough, you may notice that the choice of what to output for any given n depends only on the "equivalence class" determined by n % 15. Once we know this equivalence class we know whether n is divisible by 3 and whether n is divisible by 5.

For example, consider the members of the equivalence class [6], those numbers that have a remainder of 6 when divided by 15.

```
6, 21, 36, 51, 66, 81, 96
```

Plainly none of these numbers is divisible by 15; each is 6 more than a multiple of 15. But each of these numbers is divisible by 3, since every member of this equivalence class can be written as

```
15 * n + 6
```

for some n, and each of the two terms in the sum is divisible by 3.

That is, if a number belongs to this equivalence class, its correct output is "fizz".

We can do something similar for the other 14 equivalence classes, which you can see if we cleverly annotate CYCLE_OF_15:

```
CYCLE_OF_15 = ['fizzbuzz',   #  0, 15, 30, 45, 60, 75,  90
               None,         #  1, 16, 31, 46, 61, 76,  91
               None,         #  2, 17, 32, 47, 62, 77,  92
               'fizz',       #  3, 18, 33, 48, 63, 78,  93,
               None,         #  4, 19, 34, 49, 64, 79,  94
               'buzz',       #  5, 20, 35, 50, 65, 80,  95
               'fizz',       #  6, 21, 36, 51, 66, 81,  96
               None,         #  7, 22, 37, 52, 67, 82,  97
               None,         #  8, 23, 38, 53, 68, 83,  98
               'fizz',       #  9, 24, 39, 54, 69, 84,  99
               'buzz',       # 10, 25, 40, 55, 70, 85, 100
               None,         # 11, 26, 41, 56, 71, 86
               'fizz',       # 12, 27, 42, 57, 72, 87
               None,         # 13, 28, 43, 58, 73, 88
               None]         # 14, 29, 44, 59, 74, 89
```

In fact, we can rewrite the previous "if / elif" solution in terms of equivalence classes, which possibly makes it clearer what's happening:

```python
def fizz_buzz(n: int) -> str:
    equivalence_class = n % 15

    # "divisible by 3 but not by 15" classes
    if equivalence_class in [3, 6, 9, 12]:
        return 'fizz'

    # "divisible by 5 but not by 15" classes
    elif equivalence_class in [5, 10]:
        return 'buzz'

    # "divisible by 15" class
    elif equivalence_class == 0:
        return 'fizzbuzz'

    else:
        return str(n)
```

This version also doesn't depend on the order in which we check the equivalence classes.

Another way to see this pattern is to print out the results in groups of 15:

```python
# remap to shorter
remap = {'fizzbuzz': 'fb', 'fizz': 'f', 'buzz': 'b'}

output = [remap.get(fizz_buzz(n), '-') for n in range(1, 101)]

for start in range(0, 100, 15):
    cycle = output[start:start+15]
    print(" ".join(cycle))
```

The result is

```
- - f - b f - - f b - f - - fb
- - f - b f - - f b - f - - fb
- - f - b f - - f b - f - - fb
- - f - b f - - f b - f - - fb
- - f - b f - - f b - f - - fb
- - f - b f - - f b - f - - fb
- - f - b f - - f b
```

which clearly shows the pattern.

None as a Sentinel

One underlying tension throughout the solutions in this book is between my framing that there are four possible outputs (which is true if you think of "as-is" as an output) and the reality that there are infinitely many possible outputs (which is true if you think of "1" and "2" and "4" as outputs).

If the input is 1 you need to look at CYCLE_OF_15[1] and somehow get the correct output of "1". And if the input is 16 you also need to look at CYCLE_OF_15[1] but this time return a correct output of "16". Obviously a single list can't contain two different numbers at position 1.

One possibility would be to create the list dynamically:

```python
from typing import List

def cycle_of_15(n: int) -> List[int]:
    asis = str(n)
    return ['fizzbuzz', asis, asis, 'fizz', asis, 'buzz', 'fizz',
            asis, asis, 'fizz', 'buzz', asis, 'fizz', asis, asis]

def fizz_buzz(n: int) -> str:
    return cycle_of_15(n)[n % 15]
```

This feels "wasteful" somehow, creating what's essentially the same list each time you call the function. But also there's a sense that what's important about the list is constant, and that dynamically generating the list obscures that.

That leaves the question of how to represent "as is" in the list. In Python we often use None to represent "no value". (Many other programming languages have a similar but not identical null value.)

So here you could think of None in CYCLE_OF_15 as representing "no special value", which is another way of saying "as is".

Sometimes you need a sentinel value that's different from None, for example in the following:

```python
def greet(name: str, title: str) -> str: ...
```

Imagine that we would like to handle two separate cases: one where no title is supplied, another whether the caller explicitly passes in None for the title.

In that case we'll sometimes create a sentinel object which (necessarily) only equals itself.

```python
DEFAULT_TITLE = object()

def greet(name: str, title = DEFAULT_TITLE) -> str:
    if title == DEFAULT_TITLE:
        # caller did not provide a title
        return f"{name} the {name}-iest"
    elif title is None:
        # caller explicitly provided None
        return f"{name}
    else:
        return f"{title} {name}"
```

This is not a particularly common thing to do, but occasionally it's useful.

Similarly we could have defined

```
AS_IS = object()

CYCLE_OF_15 = [
    'fizzbuzz', AS_IS, AS_IS, 'fizz', AS_IS, 'buzz', 'fizz',
    AS_IS, AS_IS, 'fizz', 'buzz', AS_IS, 'fizz', AS_IS, AS_IS
]
```

although then the logic inside our fizz_buzz function would have been more complicated. Probably this would not have been a good tradeoff.

Truthiness and Logic

Recall our fizz_buzz function:

```
def fizz_buzz(n: int) -> str:
    return CYCLE_OF_15[n % 15] or str(n)
```

What is that or doing in there? You are likely familiar with using and and or in a logic setting:

```
if n % 3 == 0 or n % 5 == 0:
    print("either fizz or buzz or fizzbuzz")
else:
    print("as is")
```

Hopefully it's somewhat obvious what and and or do when you give them boolean values (that is, True and False values).

In Python we tend to use any value where a boolean is expected. Some values are "truthy" and behave like True. Some values are "falsy" and behave like False. Typically something is falsy if it's

- False

- None
- 0
- '' (the empty string)
- an empty container (e.g. [] or {})

Most other things are truthy. Accordingly, you can use a list as boolean to check whether it's empty:

```python
def average(xs: List[float]) -> float:
    if xs:
        return sum(xs) / len(xs)
    else:
        raise ValueError("cannot take average of an empty list")
```

This sort of idiom is extremely common in Python code.

We already know how and and or work with boolean values:

```python
# is True if both are True, False otherwise
divisible_by_15 = x % 3 == 0 and x % 5 == 0

# is True if either is True, False otherwise
not_as_is = x % 3 == 0 or x % 5 == 0
```

But what do they do with merely truthy or falsy values?

```python
# ?
empty_or_full = [] or [1, 2, 3]
empty_and_full = [] and [1, 2, 3]
full_or_empty = [1, 2, 3] or []
full_and_empty = [1, 2, 3] and []
```

Try it yourself. It turns out that if a is truthy, a or b returns a, and if a is falsy, a or b returns b. Likewise, if a is truthy, a and b returns b, and if a is falsy, a and b returns a.

First, you should check that when a and b are booleans this gives the answer you expect. And then you should check that when a and b are merely truthy or falsy that this gives you the truthiness you expect.

For example, [1, 2, 3] and [1, 2] equals [1, 2], which is truthy. And [1, 2, 3] or [1, 2] equals [1, 2, 3], which is also truthy.

In particular, None is falsy, which means that

```
None or s
```

is equal to s. We made use of this in our solution:

```python
def fizz_buzz(n: int) -> str:
    return CYCLE_OF_15[n % 15] or str(n)
```

If n is not divisible by 3 or by 5, CYCLE_OF_15[n % 15] is None, and so the result of the or is str(n). If n is divisible by 3 or by 5, then CYCLE_OF_15[n % 15] is a non-empty string, which is truthy, and so the result of the or is that string.

This suggests that we could have used False or '' or even 0 as the placeholder for "as is" in CYCLE_OF_15, but for me None seems like a better choice, as it sort of suggests "not fizz or buzz or fizzbuzz".

dict.get

Storing all those Nones is kind of clunky, but using a list forces our hand. An alternative approach is to use a dict that stores only the interesting classes:

```
DICT_OF_15 = {3: 'fizz', 6: 'fizz', 9: 'fizz', 12: 'fizz',
              5: 'buzz', 10: 'buzz',
              0: 'fizzbuzz'}

def fizz_buzz(n: int) -> str:
    return DICT_OF_15.get(n % 15, str(n))
```

Python's `dict.get` is a safe way of looking up keys that may or may not exist in the dictionary. Here we provide `str(n)` as the default value:

```
DICT_OF_15.get(n % 15, str(n))
```

If `n % 15` is a key in the `dict`, we return its corresponding value; otherwise we return `str(n)`. If you don't supply a default value, Python will use `None` as the default value, which means we also could have written

```
DICT_OF_15.get(n % 15) or str(n)
```

which is more similar to the original list-based solution.

4. Euclid's Solution

Euclid was an ancient Greek mathematician sometimes called the father of geometry. He predated Fizz Buzz by thousands of years, but this is how I like to think he would have solved it:

```python
def fizz_buzz(n: int) -> str:
    hi, lo = max(n, 15), min(n, 15)

    while hi % lo > 0:
        hi, lo = lo, hi % lo

    return {1: str(n), 3: "fizz", 5: "buzz", 15: "fizzbuzz"}[lo]
```

In this chapter we'll explore why this works and why Euclid might have solved it this way.

Prime Numbers

A positive number is *prime* if it cannot be written as the product of two smaller numbers.

3 and 5 are both prime numbers, since there's no way to write them as such a product. 15 is not a prime number, as we can write it as 3 * 5. By convention, 1 is also not a prime number, since we can only write 1 * 1.

One way to check whether a number is prime by trying to divide it by every number smaller than itself, starting at 2:

```python
def is_prime(n: int) -> bool:
    """
    n is prime if it's at least 2 and if it's not
    divisible by any smaller number (other than 1)
    """
    return (
        n >= 2 and
        all(n % d > 0 for d in range(2, n))
    )


assert all(is_prime(n) for n in [2, 3, 7, 11, 83, 89, 97])
assert not any(is_prime(n) for n in [4, 50, 91])
```

In order to check whether some number n is prime, this function has to take (in the worst case) approximately n actions: first a check that n is at least two, and then a divisibility check for each of 2, 3, ... n - 1. (In the best case it will take a lot fewer actions; for example, is_prime(1000) will stop as soon as it computes 1000 % 2 == 0.)

This means that if we wanted a list of the *all* the primes up to n, a worst case estimate is that it would take approximately n * n actions.

We can do a little better, though. We really only need to check for divisors up to math.sqrt(n), since if a and b are both larger than math.sqrt(n) then their product is larger than n, and in particular is not n:

```python
import math


def int_sqrt(n: int) -> int:
    return int(math.sqrt(n))


def is_prime2(n: int) -> bool:
    return (
        n >= 2 and
        all(n % d > 0 for d in range(2, int_sqrt(n) + 1))
    )
```

Let's see how slow this is:

```
from typing import List

def primes_up_to(n: int) -> List[int]:
    return [i for i in range(2, n + 1) if is_prime2(i)]
```

As always, we write a couple of test cases:

```
assert primes_up_to(20) == [2, 3, 5, 7, 11, 13, 17, 19]
assert primes_up_to(100)[-3:] == [83, 89, 97]
```

Then, if you're using IPython you can use the %timeit magic to see how this scales:

```
%timeit primes = primes_up_to(100)
```

On my laptop this results in

```
100          .077 ms
1000        1.051 ms (14x the previous)
10000       13.4  ms (13x)
100000   209      ms (16x)
1000000  4860     ms (23x)
```

That is, it takes about 5 seconds to find all the primes up to 1 million. Given the way it's scaling I don't particularly want to check 10 million and beyond.

A more efficient way is to use a trick called the "sieve", which is based on the observation that if a number is not prime than it's necessarily divisible by a smaller *prime* number.

Start with the numbers 2, ..., n as "candidate" primes. The smallest element (that is, 2) must be prime. Remove it (and remember it) and then eliminate all further multiples of 2 as not prime. The new smallest element 3 is now our next prime. Remove it and then eliminate all the multiples of 3 as not prime. Keep repeating this process. At each step the new smallest element was not divisible by any smaller prime, hence must be prime itself.

```
def primes_up_to(n: int) -> List[int]:
    candidates = range(2, n + 1)
    primes = []

    while candidates:
        # The smallest remaining number must be prime,
        # because it wasn't divisible by any smaller prime.
        p = candidates[0]
        primes.append(p)

        # Remove further multiples of p as not-prime
        candidates = [n for n in candidates if n % p > 0]

    return primes
```

And we'll repeat the test cases:

```
assert primes_up_to(20) == [2, 3, 5, 7, 11, 13, 17, 19]
assert primes_up_to(100)[-3:] == [83, 89, 97]
```

But it turns out that this version doesn't scale well either! Using %timeit again I get the following:

```
100          0.0337 ms
1000         0.764  ms   (22x the previous)
10000       45.9    ms   (60x)
100000    2590      ms   (56x)
```

That is, it takes over 2.5 seconds to find all the primes up to 100,000. This is even worse than the "slow" version! The sieve was supposed to be fast. What did we do wrong?

Performance Optimization

We can use a tool called cProfile to investigate where our code is spending its time:

```
import cProfile
cProfile.run('primes_up_to(100_000)')
```

The output (severely edited to fit in the book) looks sort of like

```
ncalls   tottime   function
1        0.078     <ipython-input-18-7da95fe8b6cd>:1(primes_up_to)
9592     2.138     <ipython-input-18-7da95fe8b6cd>:12(<listcomp>)
1        0.000     <string>:1(<module>)
1        0.000     {built-in method builtins.exec}
9592     0.000     {method 'append' of 'list' objects}
1        0.000     {method 'disable' of '_lsprof.Profiler' objects}
```

Almost all of the time is spent in `<listcomp>`; that is, doing 9600 list comprehensions.

This makes sense. Recreating the list of candidates each time by inspecting every element is overkill. When we are doing the sieve for p = 5, there's no need to check 7, since it's not a multiple of 5. But in order to recreate the list of candidates we have to iterate over every element.

How can we eliminate this? An alternative approach makes the following changes:

1. We maintain a list `candidates` of booleans of length n + 1. `candidates[i]` is `True` if we haven't yet ruled out `i` as a prime number. We will systematically go through the list and set every composite (not prime) number to `False`.
2. We check every candidate from 2 to `math.sqrt(n)`. This is sufficient because if some `m <= n` is composite and can be written as `a * b` then at least one of a and b must be smaller than `sqrt(n)`, and we'll set m to `False` when considering that number (or one of its prime factors).
3. If a candidate p has already been eliminated as a candidate prime, we skip it
4. Otherwise it's a prime number. Then we eliminate p `** 2`, p `** 2 + p`, p `** 2 + 2 * p` and so on. (Smaller multiples of p will have already been eliminated by the time we get to p.)

```
def primes_up_to(n: int) -> List[int]:
    #              0       1           2, ..., n
    candidates = [False, False] + [True] * (n - 1)

    for p in range(2, int(math.sqrt(n))):
        # if we haven't already eliminated p as a prime
        if candidates[p]:
            # eliminate all multiples of p, starting at p ** 2
            for m in range(p * p, n + 1, p):
                candidates[m] = False

    # return the indices that weren't eliminated
    return [n for n, prime in enumerate(candidates) if prime]
```

Once again we can check the timings:

```
100            0.0067 ms
1000           0.0884 ms   (13x the previous)
10000          0.911  ms   (10x)
100000         9.16   ms   (10x)
1000000     139       ms   (15x)
10000000   2240       ms   (16x)
```

This version is faster for small input sizes, but more importantly it grows much more slowly as the input size increases. You can see that it finds all prime numbers up to 10,000,000 faster than the previous sieve implementation found all the prime numbers up to 100,000.

Mathematically both versions are the same algorithm, but the implementation differences turn out to be pretty crucial to performance.

Factorization

It turns out that every positive integer can be written in a unique way as a product of prime numbers. (Here "unique" means "order doesn't matter", as plainly 3 * 5 == 5 * 3 and so on.) We call this product the "prime factorization".

Using the sieve it's easy enough to find the prime factorization of a number:

```python
from typing import List

def factorize(n: int) -> List[int]:
    primes = primes_up_to(n)

    factors = []

    for p in primes:
        # p might divide n more than once
        while n % p == 0:
            factors.append(p)
            n = n // p
        # once we reach 1 there are no more prime factors
        if n == 1:
            break

    return factors
```

And as always we write a few test cases:

```python
assert factorize(15) == [3, 5]
assert factorize(150) == [2, 3, 5, 5]
assert factorize(7) == [7]
```

What Prime Factorization Has to Do with Fizz Buzz

If some number n is divisible by 3, then 3 is necessarily a prime factor of n, and vice versa. Same thing goes for 5.

So here's another "simple" solution:

```python
def fizz_buzz_factorization(n: int) -> str:
    prime_factors = factorize(n)

    if 3 in prime_factors and 5 in prime_factors:
        return "fizzbuzz"
    elif 3 in prime_factors:
        return "fizz"
    elif 5 in prime_factors:
        return "buzz"
    else:
        return str(n)
```

Why the scare quotes? Because this "simple" solution required us to first implement a prime number sieve and factorization algorithm. The solution at the beginning of the chapter certainly didn't do all that!

Greatest Common Divisors and Least Common Multiples

Frequently in mathematics we would like to know the greatest common divisor of two numbers; that is, the largest number that divides both of them. For example, we do this when we reduce fractions.

Imagine we're given the fraction 24 / 44. We calculate (don't worry about how right now) that the greatest common divisor of 24 and 44 is 4, and then we cancel out a factor of 4 from the numerator and denominator:

```
24 / 44 = (6 * 4) / (11 * 4) = 6 / 11
```

One way to find the greatest common divisor is to factorize the two numbers and take the product of all the common factors. In this case, 24 factorizes as [2, 2, 2, 3] and 44 as [2, 2, 11], so the common factors are [2, 2], and their product is 4. More generally:

```python
def gcd_factorize(n: int, m: int) -> int:
    n_factors = factorize(n)
    m_factors = factorize(m)
    gcd = 1

    # Stop when either list is empty
    while n_factors and m_factors:
        # Greatest remaining factors of both are equal
        # so multiply the gcd by that factor
        if n_factors[-1] == m_factors[-1]:
            gcd *= n_factors.pop()
            m_factors.pop()
        # Largest factor of m is not a factor of n
        elif n_factors[-1] < m_factors[-1]:
            m_factors.pop()
        # Largest factor of n is not a factor of m
        else:
            n_factors.pop()

    return gcd
```

Let's check a few cases:

```python
assert gcd(24, 44) == 4
assert gcd(44, 24) == 4
assert gcd(3, 5) == 1
assert gcd(100, 10) == 10
```

It would probably feel more natural to go through the factors in the order they're given (that is, front to back). We start from the end because `pop()`-ing the last element off a list is a very cheap operation; removing the first element requires (in essence) making a copy of the list.

With the sizes of lists we're dealing with it doesn't really matter, but it's good practice to get in the habit of using efficient idioms and avoiding inefficient ones. (This will be a common theme throughout this book.)

Sometimes we also care about the *least common multiple* of two numbers; that is, the smallest number that is itself divisible by both. For example, we use this to find a common denominator when adding fractions:

```
3 / 10 + 3 / 4 = 6 / 20 + 15 / 20
              = 21 / 20
```

Here we used the fact that the least common multiple of 10 and 4 is 20.

We were able to compute the gcd of two numbers by taking the intersection of their prime factors. We can similarly find the lcm by taking the union of their prime factors:

```python
def lcm_factorize(n: int, m: int) -> int:
    n_factors = factorize(n)
    m_factors = factorize(m)
    lcm = 1

    # Stop when both lists are empty
    while n_factors or m_factors:
        # no more prime factors of n
        if not n_factors:
            lcm *= m_factors.pop()
        # no more prime factors of m
        elif not m_factors:
            lcm *= n_factors.pop()
        # same largest prime factor, only use it once
        elif n_factors[-1] == m_factors[-1]:
            lcm *= n_factors.pop()
            m_factors.pop()
        # largest factor of m is not a factor of n
        elif n_factors[-1] < m_factors[-1]:
            lcm *= m_factors.pop()
        # largest factor of n is not a factor of m
        else:
```

```
        lcm *= n_factors.pop()

    return lcm
```

And as usual we write some tests:

```
assert lcm_factorize(10, 4) == 20
assert lcm_factorize(4, 10) == 20
assert lcm_factorize(3, 5) == 15
assert lcm_factorize(10, 100) == 100
```

One final observation: gcd(n, m) is the product of only the common factors. lcm(n, m) is the product of all the factors of both, but with common factors only counted once (even though they appear twice). And n * m is the product of all the factors of both as many times as they appear.

This means that `lcm(n, m) * gcd(n, m) == n * m`. So if you know how to compute one, you know how to compute the other. (Indeed, currently Python has a `math.gcd` function but no `math.lcm` function, although this is tentatively scheduled to change in Python 3.9.)

gcd and Fizz Buzz

What does any of this have to do with Fizz Buzz?

Well, `gcd(n, m)` gives us the product of the common factors of n and m. And the factors of 15 are `[3, 5]`.

That means that `gcd(n, 15)` is either 1 (if n has no factors of 3 or 5), 3 (if n has a factor of 3 but not a factor of 5), 5 (if n has a factor of 5 but not a factor of 3), or 15 (if n has a factor of 3 and a factor of 5).

Of course, saying "n has a factor of 3" is the same as saying "n is divisible by 3". Now this is starting to look like Fizz Buzz.

Using Python's `math.gcd` function, you could just do:

```
import math

def fizz_buzz(n: int) -> str:
    choices = {1: str(n), 3: 'fizz', 5: 'buzz', 15: 'fizzbuzz'}
    return choices[math.gcd(n, 15)]
```

So that's the idea behind our solution. But that wasn't our solution.

Euclid's Algorithm

How would we go about computing the greatest common divisor of two numbers without computing their prime factorizations first? (Obviously we can use math.gcd, but what if we didn't have that?)

The brute force way is to take all numbers between 1 and the smaller number, keep only the ones that divide both of the numbers, and take the largest:

```
def gcd_slow(n: int, m: int) -> int:
    return max(i
               for i in range(1, min(n, m) + 1)
               if n % i == m % i == 0)
```

If n and m are both large, that's a lot of numbers to check. For example, to compute

```
gcd_slow(97039801, 97313179)
```

on my laptop takes about 8 seconds. (Those are both prime numbers, so their gcd is 1.) You could also check from highest to lowest and stop when you find one, but that would still take a long time when the gcd is small.

A more efficient way to compute the gcd is using *Euclid's algorithm*. You remember Euclid from the start of the chapter.

Say we have two numbers m `<=` n with greatest common divisor d. Then we can do "long division" and write n `=` c `*` m `+` r where 0 `<=` r `<` m. That is, c `==` n `//` m is the result of integer division, and r `==` n `%` m is the remainder.

I claim that d `==` gcd(m, n) `==` gcd(r, m). That is, the greatest common divisor of m and n is the same as the greatest common denominator of m and n `%` m. Why is this?

Well, if x is any number that divides both m and n, then c `*` m is also divisible by x (since m is), which means that r `=` n `-` c `*` m must be divisible by x.

On the other hand, if y is any number that divides both m and r, then c `*` m is also divisible by y (since m is), which means that n `=` c `*` m `+` r must be divisible by y.

That is, {common divisors of m and n} and {common divisors of m and r} are the exact same sets of numbers. Any number in the set on the left is also in the set on the right, and any number in the set on the right is also in the set on the left.

In particular, they have the same greatest element. The greatest element of the set on the left is (by definition) gcd(m, n), and the greatest element of the set on the right is (by definition) gcd(m, r), so these gcds must be the same.

So now imagine we are trying to compute gcd(n, m) with n `>=` m. We just showed that this is is the same as gcd(m, n % m). If n `%` m `==` 0, then this equals m, and we're done. In particular, if m `==` n, this is the case. Otherwise, necessarily m `<` n and we repeat.

At each step, either the larger of the pair gets smaller, or we're done.

This leads to a much faster implementation:

```python
def gcd(n: int, m: int) -> int:
    # Want n >= m
    n, m = max(m, n), min(m, n)

    # gcd(n, m) = gcd(m, n % m)
    while n % m > 0:
        n, m = m, n % m

    # When n % m == 0, n is a multiple of m, so m is the gcd
    return m
```

For the same two large prime numbers this takes about 1/1000 of a second.

"Euclid's solution" was just an explicit invocation of this algorithm with m = 15. That is, it was an extremely opaque (but fast) computation of gcd(n, 15), which allows us to choose the correct Fizz Buzz result.

5. Trigonometry

You remember trigonometry from high school, right?

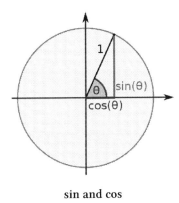

sin and cos

Here we'll use it to solve Fizz Buzz:

```python
import math

def fizz_buzz(n: int) -> str:
    fizz = 'fizz' * int(math.cos(n * math.tau / 3))
    buzz = 'buzz' * int(math.cos(n * math.tau / 5))
    return (fizz + buzz) or str(n)
```

And they said trigonometry would never be useful!

Sin and Cos

Imagine a point on the unit circle, as pictured above. If the point makes an angle of theta with the origin, we define sin(theta) to be the point's y-coordinate and cos(theta) to be the point's x-coordinate.

Although you may be used to measuring angles in *degrees* (in which a right angle represents 90 degrees and a full rotation represents 360 degrees), mathematicians prefer to represent angles in *radians*, in which a full rotation represents 2 * math.pi radians:

```python
def degrees_to_radians(degrees: float) -> float:
    rotations = degrees / 360
    return rotations * 2 * math.pi

# Half a rotation is math.pi radians or about 3.14
assert 3.14159 < degrees_to_radians(180) < 3.14160
```

```python
def radians_to_degrees(radians: float) -> float:
    rotations = radians / 2 / math.pi
    return rotations * 360

assert 179.99 < radians_to_degrees(3.141592653589793) < 180.01
```

You probably know that math.pi is approximately 3.14 which means that a full rotation is about 6.28 radians.

A rotation of 2 * math.pi (or 360 degrees) represents exactly once around the circle, which means that the point on the unit circle with angle theta + 2 * math.pi is exactly the same as the point on the unit circle with angle theta. In particular, that means that for any theta we have

```python
math.sin(theta + 2 * math.pi) == math.sin(theta)
math.cos(theta + 2 * math.pi) == math.cos(theta)
```

In other words, these functions repeat every 2 * math.pi radians.

Notice also that if a point is on the unit circle, its x-coordinate satisfies -1 <= x <= 1, and its y-coordinate satisfies -1 <= y <= 1. So math.sin and math.cos satisfy the same property.

Finally, notice that cos(theta) (which is the x-coordinate) equals 1 when theta is 0 and again at 2 * math.pi, 4 * math.pi and so on. It equals -1 when theta is math.pi (180 degrees) and again at 3 * math.pi, 5 * math.pi, and so on. For all other values it is strictly between -1 and 1.

These will all end up being important to our solution.

int

The int function converts things into integers. How does it convert them? Well, it depends on the thing.

If you give it no arguments, it returns 0:

```
assert int() == 0
```

If you give it an integer, it returns that integer:

```
assert int(12) == 12
assert int(-5) == -5
```

If you give it a string representing an integer, it does the right thing:

```
assert int("12") == 12
assert int("-5") == -5
```

If you give it a float, it truncates toward 0:

```
assert int(-1.1) == -1
assert int(-1.0) == -1
assert int(-0.9) == 0
assert int(0.9) == 0
assert int(1.0) == 1
assert int(1.1) == 1
```

(However, if you give it a *string* representing a float, like "1.2", it raises a ValueError.)

Now imagine that we have some number x that we know is between -1 and 1. Then int(x) does the following:

```
def restricted_int(x: float) -> int:
    assert -1 <= x <= 1, "we assumed this"

    if x == -1:
        return -1
    elif x == 1:
        return 1
    else:
        # -1 < x < 1, so "truncate toward 0" returns 0
        return 0
```

This is the sense in which we use int in our solution.

Pi and Tau

The Tau Manifesto argues that math.pi, representing half a rotation, is the wrong circle constant. It advocates that we should instead use math.tau == 2 * math.pi as the circle constant.

I won't delve into the manifesto's arguments (you can read them yourself); however, it turns out that math.tau will also make our solution simpler, and so we'll make use of it ourselves to get rid of the pesky factor of 2 that keeps cropping up.

Using math.tau, the period of repetition is much cleaner:

```
math.sin(theta + math.tau) == math.sin(theta)
math.cos(theta + math.tau) == math.cos(theta)
```

And now we have that cos(theta) is 1 precisely when theta is a multiple of math.tau, and that it's -1 precisely when theta is math.tau / 2, math.tau * 3 / 2, and so on.

Given the discussion in the previous section, that means that int(cos(theta)) is 1 if theta is a multiple of math.tau, is -1 if theta is half a multiple of math.tau, and is 0 otherwise.

That is, it behaves like

```
def int_cos(theta: float) -> int:
    rotations = theta / math.tau

    if rotations % 1 == 0:
        # Some number of full rotations
        return 1
    elif rotations % 1 == 0.5:
        # Half a rotation
        return -1
    else:
        return 0
```

So now let's think about our trigonometric function:

```
int(math.cos(n * math.tau / 3))
```

If n is a multiple of 3, then n * math.tau / 3 is the angle corresponding to some number of full rotations. For example, if n is 3, then it just equals math.tau, or one full rotation. If n is 6, it equals 2 * math.tau, or two full rotations. And so on.

That is, when n is a multiple of 3, the value of math.cos is 1, and since int(1) is 1, the whole expression equals 1.

If n is not a multiple of 3, then this represents neither a full rotation or a half rotation. (For example, if n is 1, it's 1/3 of a rotation, and if n is 2 it's 2/3 of a rotation.) This means that cos results in a number strictly between -1 and 1, and that calling int on it truncates it to zero.

To sum up,

```
int(math.cos(n * math.tau / 3))
```

equals 1 when n is a multiple of 3, and it equals 0 otherwise. In other words (thinking of 1 as True and 0 as False) this expression is simply calculating "is n divisible by 3".

Similar reasoning reveals that

```
int(math.cos(n * math.tau / 5))
```

equals 1 if n is a multiple of 5 and 0 otherwise. This is starting to get us close to a solution.

Operator Overload

It's pretty obvious what 2 + 3 means:

```
assert 2 + 3 == 5
```

Python overloads the + operator to work in other circumstances too:

```
# lists
assert [1, 2] + [3, 4] == [1, 2, 3, 4]

# tuples
assert (1, 2, 3) + (4, ) == (1, 2, 3, 4)

# strings
assert "12" + "34" == "1234"
```

Multiplication is similar. It's pretty obvious what 2 * 3 means:

```
assert 2 * 3 == 6
```

But we can also do:

```
# lists
assert [0] * 3 == [0, 0, 0]

# tuples
assert (1, 2) * 4 == (1, 2, 1, 2, 1, 2, 1, 2)

# strings
assert "waka" * 2 == "wakawaka"
```

We can use this to our benefit in the following way:

```
# string * 0 is an empty string
assert 'fizz' * 0 == ''

# string * 1 is itself
assert 'fizz' * 1 == 'fizz'
```

In the previous section we saw that

```
int(math.cos(n * math.tau / 3))
```

equals 1 if n is a multiple of 3 and 0 otherwise. This means that

```
fizz = 'fizz' * int(math.cos(n * math.tau / 3))
```

equals `'fizz'` if n is a multiple of 3 and `''` otherwise.
Similarly,

```
buzz = 'buzz' * int(math.cos(n * math.tau / 5))
```

equals `'buzz'` if n is a multiple of 5 and `''` otherwise.

Then `fizz + buzz` concatenates those strings, which results in

- `'fizzbuzz'` if n is a multiple of both 3 and 5 (i.e. divisible by 15)
- `'fizz'` if n is divisible by 3 but not by 5
- `'buzz'` if n is divisible by 5 but not by 3
- `''` if n is not divisible by either

Only the last is falsy, which means that

```
(fizz + buzz) or str(n)
```

gives us the answer we want.

6. A Big Number

I hope you like big numbers:

```python
import re

big_number = 0x134591c9a2c8e4d1647268b23934591c9a2c8e4d16

chunks = re.findall('(0|10|110|111)', f"{big_number:0>167b}")

def label(chunk: str, n: int) -> str:
    labels = [str(n), '', 'fizz', '', '', '', 'buzz', 'fizzbuzz']
    return labels[int(chunk, 2)]

output = [label(chunk, n+1) for n, chunk in enumerate(chunks)]
```

In this chapter we'll explore where this big number came from and how it solves Fizz Buzz.

Decimal, Binary, and Hexadecimal

Normally we write numbers in *decimal* format, or base 10. There are 10 digits, 0 to 9, the rightmost of which represents the number of 1s, the next the number of 10s, the next 100s, and so on.

```
from typing import List

def decimal_digits(n: int) -> List[int]:
    """
    Convert the number n to a list of its digits.
    """
    digits: List[int] = []

    # If the number is 0, it has no more digits.
    while n > 0:
        # The rightmost digit is n % 10
        digits.append(n % 10)
        # Then replace n with n // 10 and repeat
        n = n // 10

    digits.reverse()
    return digits
```

This function ignores leading zeroes. Accordingly, it reports that 0 has "no digits":

```
assert decimal_digits(0) == []
assert decimal_digits(10) == [1, 0]
assert decimal_digits(123) == [1, 2, 3]
```

Let's think through what it's doing for the number 123. We find the rightmost digit 123 % 10 == 3 and then throw it away to get 123 // 10 == 12. The next rightmost digit is then 12 % 10 == 2, which we throw away to get 12 // 10 == 1. The final digit is 1 % 10 == 1, which we throw away to get 1 // 10 == 0, at which point we stop.

Now our list of digits is [3, 2, 1], so we have to reverse it and then it's the right answer.

We can reconstruct a number from its digits by performing the same process in reverse:

```python
def from_decimal_digits(digits: List[int]) -> int:
    # start with 0
    n = 0

    for digit in digits:
        # multiply by 10 to make "room" for the digit
        n = n * 10
        # and then add it
        n += digit

    return n
```

Which we test as usual:

```python
assert from_decimal_digits([]) == 0
assert from_decimal_digits([1, 0]) == 10
assert from_decimal_digits([1, 2, 3]) == 123
```

You should convince yourself that this logic is correct.

Of course, there is nothing particularly special about the number 10, except that most of us have 10 fingers. Math works perfectly well if you use base 9, or base 8, or even base 2.

In fact, behind the scenes computers mostly use base 2 ("binary"), in which the only digits are 0 or 1. In that case the rightmost digit is the 1s place, the next digit the 2s place, the next the 4s place, and so on.

So the binary number with digits `[1, 0, 0, 1, 1]` represents $16 + 2 + 1 = 19$.

In Python you can write binary literals by prefixing them with `0b`, and you can get binary representations of numbers using `bin()`:

```python
assert 0b10011 == 19
assert bin(19) == '0b10011'   # a string
```

You can also use `int()` and specify the base explicitly:

```
assert int('10011', base=2) == 19
```

It's not hard to modify our decimal_digits functions to use binary representations, and I encourage you to do so.

One other somewhat common representation used by computers is base-16, or *hexadecimal*. As we run out of numeric digits before we get to 16, in hexadecimal we use a for 10, b for 11, c for 12, d for 13, e for 14, and f for 15.

We can write hexadecimal literals by prefixing them with 0x, and we can convert integers to hexadecimal using hex():

```
# 0xf83 is 15 * 256 + 8 * 16 + 3 * 1 == 3971
assert 0xf83 == 3971
assert hex(3971) == '0xf83'  # a string
assert int('f83', base=16) == 3971
```

Notice that large hexadecimal numbers are *shorter* than their decimal equivalents (and large binary numbers *longer* than their decimal equivalents) since we can fit in more numbers before having to add an extra digit.

A Secret Encoding

In Chapter 1, we "solved" the problem by explicitly putting the results in a list:

```
FIZZ_BUZZ = [1, 2, 'fizz', ...]
```

A less obvious way of doing the same would be to somehow *encode* the results in a way so that they didn't look like the results. As we've seen repeatedly, there are basically 4 different output cases: as-is, "fizz", "buzz" and "fizzbuzz". There are also four different two-digit binary numbers. Thus, we could encode each case in binary:

```python
def encode(n: int) -> str:
    if n % 15 == 0:
        # binary '3' = fizzbuzz
        return '11'
    elif n % 5 == 0:
        # binary '2' = buzz
        return '10'
    elif n % 3 == 0:
        # binary '1' = fizz
        return '01'
    else:
        # binary '0' = as-is
        return '00'
```

And have something like

```python
# ['00', '00', '01', ...]
fizz_buzz_encoded = [encode(n) for n in range(1, 101)]
```

Which we could squeeze together to represent a single binary number:

```python
# '000001...'
fizz_buzz_binary = ''.join(fizz_buzz_encoded)
```

Which represents an actual 59-digit number:

```python
# 28648571938333232209340275945476261587143340067117300451149494
fizz_buzz_num = int(fizz_buzz_binary, base=2)
```

And then we can save a few digits (and be more cryptic) by writing it in hexadecimal:

```
# '0x490610c12418430490610c12418430490610c124184304906'
hex(fizz_buzz_num)
```

That big number is just an obfuscated way of representing all the precomputed answers. But that's not the big number we used (ours was smaller).

Prefix Codes

The encoding in the previous section used two binary digits to encode each of the four classes. But the classes are not equally likely. Using the gcd as a stand-in for the classes (mostly because it's the least work to compute), let's count them:

```
from collections import Counter
import math

gcds = [math.gcd(n, 15) for n in range(1, 101)]
counts = Counter(gcds)

assert counts == {1: 53, 3: 27, 5: 14, 15: 6}
```

More than half of the time we have gcd(n, 15) == 1, the "as is" case. What if we could use a shorter encoding for that case?

For instance, what if we used 0 to encode "as-is"? The challenge is that if we do that then every other encoding has to start with 1. Otherwise, if we saw a message starting with 0 we wouldn't know whether it was "as-is" or the start of some other encoding.

What we need is a *prefix* code, where no codeword is a prefix of any other code word. If some class is encoded as 10, then we don't want any other encoding to start with 10. That way, when we see 10 we know for certain it represents the "10" class and not the start of a different class.

Huffman Coding

One well-known prefix code is the Huffman code, which gives the shortest way of encoding symbols with a known probability distribution (which is what we need to do).

We find the code by creating a *tree* of our classes, using the following algorithm:

- Create a leaf node for each class, with associated count.
- As long as we have more than one node, take the two nodes with lowest counts and merge them, making one of them the "0" child and one of them the "1" child.
- Then the path to each class represents its encoding.

In our case, we'll end up with a tree that looks like the following:

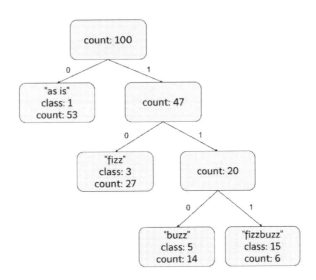

huffman tree

Reading off the encodings from the paths, you can see that we'll have

```
0 -> as-is
10 -> fizz
110 -> buzz
111 -> fizzbuzz
```

So now let's figure out how to create this tree.

Data Modeling

The leaf nodes in our tree need to know

- which class they represent, and
- the associated count

The merged nodes in our tree need to know

- who their child nodes are, and
- the total count

We need to decide how to represent these nodes in Python.

Why Not Tuples?

One way we could represent nodes is as tuples. We could represent a leaf node as a pair (count, class) and a merged node as a pair (count, children).

```
leaf_node = (53, 1)
```

This is a nice, compact representation. It's really easy to put things into tuples. However, I prefer not to model my data this way.

One reason is that the data itself gives you no indication of what it represents. What's 53? What's 1?

Relatedly, the only way to access elements is by either numeric index:

```
count = leaf_node[0]
```

or by unpacking:

```
count, cla55 = leaf_node
```

This requires a lot of mental energy to remember which field is in which position, and it makes the code a lot less readable.

Why Not Dicts?

Many people use dicts to represent structured data:

```
leaf_node = {"count": 53, "class", 1}
```

This is also not my preferred approach. One reason is that I don't like using string keys to access known values. It's awkward to type and error-prone, and your IDE can't autocomplete the field names.

```
count = leaf_node["count"]
cla55 = leaf_node["cla55"]  # oops, this is an error
```

I also don't like that they're *mutable*. There are occasions when you need to be able to modify your data, but they're much rarer than you'd think.

Finally, I don't like that they're untyped, although starting in Python 3.8 there's a `TypedDict` that addresses this.

Why Not Classes?

Classes involve a lot of boilerplate:

```
class LeafNode:
    def __init__(self, count: int, cla55: int) -> None:
        self.count = count
        self.cla55 = cla55


leaf_node = LeafNode(53, 1)
```

And they involve overhead (each instance takes up more memory than a tuple would). And they're mutable. And you'd have to add more boilerplate if you wanted them to print nicely:

```
# <__main__.LeafNode object at 0x7feffc88ae50>
print(leaf_node)
```

or compare nicely:

```
# Different Python objects, so not equal
assert not leaf_node == LeafNode(53, 1)
```

One positive is that you can just "dot in" to access elements:

```
assert leaf_node.count == 53
```

And your text editor or IDE can autocomplete these accesses, and flag errors in them.

Why Not DataClasses

Python 3.7 introduced dataclasses, which eliminate some of the boilerplate for classes that are mostly intended as "data holders", in exchange for having to use a decorator:

```
from dataclasses import dataclass

@dataclass
class LeafNode:
    count: int
    cla55: int

leaf_node = LeafNode(53, 1)
```

The @dataclass decorator provides nice default string representations and equality checks, and if you provide an extra order=True argument then you get order comparisons by field as well:

```
@dataclass(order=True)
class LeafNode:
    count: int
    cla55: int

assert LeafNode(27, 3) < LeafNode(53, 1)
```

Other than that they mostly have all the pros and cons of regular classes.

Why NamedTuples?

A NamedTuple is a tuple (so it's compact and immutable and typed and can be unpacked) but with named fields that you can access.

```
from typing import NamedTuple

class LeafNode(NamedTuple):
    count: int
    cla55: int

leaf_node = LeafNode(53, 1)
```

Since it's secretly a tuple, you can also compare for inequality:

```
assert LeafNode(27, 3) < LeafNode(53, 1)
```

which means that you can sort them. (This will be important soon.)

It's also a class, so you can add extra methods and properties to it.

Basically, it's the best of all possible worlds, unless you really need to mutate your data.

Why Not Sum Types?

Ideally what you'd like to do is to say something like "a Node is either a LeafNode(count, cla55) or a MergedNode(count, children)." And then you could write functions that take Nodes as input and use some kind of pattern matching to figure out which kind of node they're dealing with and handle them accordingly. Languages like Haskell and Scala encourage this kind of data modeling.

Python doesn't really support sum types in a non-awkward way. (There is a proposal to include pattern matching in Python 3.10, but it's too early to say what will come of it.) So what we'll do is define a single Node class that has missing values depending on what type of node it is. You could also define separate LeafNode and MergedNode classes, but then you'd end up having to check isinstance() all over the place.

Our Data Model

We'll start by defining a NamedTuple to represent a `Node` in the tree. It has a `count` that we will use to decide in what order to merge nodes. If it's a leaf node it will also have a `cla55` (we have to use 1337-speak since `class` is a reserved word in Python), whereas if it's a merged node it will also have `children`.

```python
from typing import Dict, NamedTuple, Optional

class Node(NamedTuple):
    count: int
    cla55: Optional[int] = None      # None if it's a merged node
    children: Dict[str, 'Node'] = {}  # {} if it's a leaf node

    @property
    def is_leaf(self) -> bool:
        """It's a leaf node if it has a `cla55`"""
        return self.cla55 is not None
```

We'll need to be able to `merge` two nodes, which we do by creating a new node that combines their counts and lists them as its children:

```python
def merge(n1: Node, n2: Node) -> Node:
    return Node(n1.count + n2.count, children={'1': n1, '0': n2})
```

Constructing the Code

Now we need to write a function to construct a tree (that is, a single top-level `Node`) from a dict of counts:

```python
def make_tree(counts: Dict[int, int]) -> Node:
    # Start with all leaf nodes
    nodes = [Node(count, s) for s, count in counts.items()]

    while len(nodes) > 1:
        # Pop off the two smallest nodes
        nodes.sort(reverse=True)
        node1 = nodes.pop()
        node2 = nodes.pop()

        # Merge them and add them back
        nodes.append(merge(node1, node2))

    return nodes[0]
```

If I call this on the gcd counts, I get the following (after formatting it nicely and deleting default values):

```
Node(count=100, children={
    '1': Node(count=47, children={
        '1': Node(count=20, children={
            '1': Node(count=6, cla55=15),
            '0': Node(count=14, cla55=5)}),
        '0': Node(count=27, cla55=3)}),
    '0': Node(count=53, cla55=1)})
```

From that we can read off the encodings as the paths to each class: '111' -> 15, '110' -> 5, '10' -> 3, '0' -> 1.

Of course, we'd rather use code to construct this mapping. Since we're dealing with a recursive data structure, we'll use a recursive function to generate all the paths:

```
def paths(node: Node, path_to_here: str = ''):
    if node.is_leaf:
        # yield the path to this cla55
        yield (node.cla55, path_to_here)
    else:
        # recursively generate the paths for each child,
        # adding this `step` to `path_to_here`
        for step, child in node.children.items():
            yield from paths(child, path_to_here + step)
```

And then we can construct forward and backward mappings:

```
tree = make_tree(counts)

encoder = {cla55: path for cla55, path in paths(tree)}
assert encoder == {15: '111', 5: '110', 3: '10', 1: '0'}

decoder = {path: cla55 for cla55, path in paths(tree)}
assert decoder == {'111': 15, '110': 5, '10': 3, '0': 1}
```

We can use this to generate a shorter encoding of the Fizz Buzz results:

```
# is 167 binary digits, vs 200 for the 2-digit encoding
binary_str = ''.join(encoder[n] for n in gcds)
```

Which results in a smaller big number (only 50 digits this time):

```
bignum = int(binary_str, base=2)
assert bignum == 28165702645490036728221788871744420885961791917334
```

which results in a smaller hexadecimal number as well:

```
assert hex(bignum) == '0x134591c9a2c8e4d1647268b23934591c9a2c8e4d16'
```

Digression: Mutable Default Values

We did something bad in our data model, and that's that we used a mutable default value for Node.children. Why is that bad? Because Python uses the exact same instance of the default value for every instance of Node. For things that are immutable, like strings and ints, this doesn't matter, but for things that are mutable it definitely does:

```
node1 = Node(.10, cla55=5)
node2 = Node(.20, cla55=15)

assert node1.children == node2.children == {}

node1.children["0"] = "danger!"

# node2's `children` was also modified
assert node2.children == {"0": "danger!"}

# as is the default for every future node!
node3 = Node(.30, cla55=3)
assert node3.children == {"0": "danger!"}
```

Adding an entry to node1.children also adds it to node2.children because they are the same object – the dict that gets used as the default value every time you create a Node object without supplying children. And then that modified dict also gets used for node3 and all future nodes.

For us it didn't matter since our code never modified those dictionaries, but one day you're going to make this mistake and it's going to cause you a lot of grief.

The same thing happens for function arguments:

```
def append_one(xs: List[int] = []) -> List[int]:
    # uh oh
    xs.append(1)
    return xs

assert append_one() == [1]
assert append_one() == [1, 1]
assert append_one() == [1, 1, 1]
assert append_one() == [1, 1, 1, 1]
# and so on
```

So be careful!

Digression: heapq

We did something else not good in building our tree, and that's that we used an inefficient algorithm.

At various times throughout this book we've roughly counted how many actions certain operations would take. We like it when functions perform a number of operations that's *linear* in the size of their input.

When this is the case, it takes roughly twice as many operations (and so roughly twice as long) for the function to run on inputs that are twice as large; 10 times as many on inputs 10 times as large, and so on.

Other functions perform a number of operations that is *quadratic* in the size of their inputs. Doubling the size of the input makes the function run four times as long; using an input 10 times as large makes the function run one hundred times as long. Such functions don't scale well and can't handle large inputs.

In this case, the problematic part of the code is

```
# Pop off the two smallest nodes
nodes.sort(reverse=True)
node1 = nodes.pop()
node2 = nodes.pop()
```

This is an unnecessarily expensive way of finding the two smallest nodes. Sorting n items takes on the order of n log n operations, and we're sorting n times, which means that our make_tree is worse than quadratic.

We can check this by timing how long it takes to create trees consisting of more and more nodes:

```
import random

def random_counts(n: int, seed: int = 12):
    random.seed(seed)
    return {i: random.random() for i in range(N)}
```

If you are using ipython you can then do

```
counts = random_counts(100)
```

```
%timeit make_tree(counts)
```

This scales really poorly, and even for 10000 nodes it's quite slow:

```
100                 .607 ms
1000               21.3  ms
10000            1610    ms
```

What's a faster way? Well, a *binary heap* is a data structure that's designed for relatively fast "insert", extremely fast "check minimum", and relatively fast "extract minimum". In Python it's implemented in the heapq library, which uses regular lists to store binary heaps but provides functions to interact with lists-as-heaps.

It's easy enough to modify make_tree to use a binary heap:

```
import heapq

def make_tree_heap(counts: Dict[int, int]) -> Node:
    # Convert list of nodes to binary heap in-place
    nodes = [Node(count, s) for s, count in counts.items()]
    heapq.heapify(nodes)

    while len(nodes) > 1:
        # Pop off the two smallest nodes
        node1 = heapq.heappop(nodes)
        node2 = heapq.heappop(nodes)

        # Merge them and add them back
        heapq.heappush(nodes, merge(node1, node2))

    return nodes[0]
```

Here `heapify` converts a list to a binary heap (by reordering its elements) in linear time. Each call to `heappop` and `heappush` takes roughly `log` n operations, and we're making 3 `*` n such calls, so this version scales like n `log` n, which is much less than quadratic:

```
100             .147 ms
1000            2.04 ms
10000           23.8 ms
100000          488  ms
```

For 10000 nodes it runs about 70x faster.

Our particular use case only involved 4 nodes, which means that we didn't really need to worry about these sorts of performance issues. But as I've mentioned before, writing efficient code is a good habit to get into.

Decoding

Enough digressions, let's get back to the problem. At this point we've generated a big number that encodes the first 100 Fizz Buzz outputs.

But having the big number is not enough. We also need to be able to decode it back into a solution of the problem.

Let's start by representing it as a binary number. We can do that using an f-string with a special :b format:

```
# '1001101000...'
f"{bignum:b}
```

Here we see our first problem: our encoding was supposed to start with 0010 (as-is, as-is, fizz). But numbers don't "remember" leading zeros. (This is why it didn't feel so wrong to me that our decimal_digits(0) was an empty list.) What we're seeing is the same thing that's happening in

```
x = int('00123')
assert x == 123
assert f"{x}" == '123'
```

Similarly, although we had leading zeros in our secret binary encoding, the resulting number doesn't "remember" them. This means we need to add them back when we convert to strings. We can do this by adding parameters to our format string to include padding:

```
x = 123
# pad with zeros on the left to length 5
assert f"{x:0>5}" == "00123"
# pad with zeros on the right to length 5
assert f"{x:0<5}" == "12300"
# pad with *s on the left to length 7
assert f"{x:*>7}" == "****123"
```

Here we happen to know that `len(binary_str)` was 167, which means we can just pad it to that length:

```
# '0010011010...'
f"{bignum:0>167b}"
```

We still need to turn this back into the Fizz Buzz outputs.

Recall our decoder:

```
# decoder == {'111': 15, '110': 5, '10': 3, '0': 1}
```

Because this is a prefix code, the start of the string corresponds to a unique class. We figure out what class that is, remove it from the string, and repeat:

```
from typing import List
import itertools

def chunk1(encoded: str) -> List[str]:
    chunks: List[str] = []

    while encoded:
        # Try length 1, length 2, ...
        for length in itertools.count(1):
            prefix = encoded[:length]
```

```
if prefix in decoder:
    chunks.append(prefix)
    encoded = encoded[length:]
    break  # out of for loop


return chunks
```

Which you can check finds the correct chunks:

```
chunks = chunk1(f"{bignum:0>167b}")
#                     as-is  as-is  fizz  as-is  buzz
assert chunks[:5] == [ '0',   '0',  '10',  '0',  '110']
```

"Accidentally Quadratic"

Earlier in this chapter we disdained our original implementation of make_tree for scaling (worse than) quadradically in the size of its inputs.

It turns out that our implementation of chunk1 is "accidentally quadratic". Although it doesn't look it's doing a quadratic number of operations, it secretly is.

If you count up the number of operations it seems like it should scale like the length of the string: start at 0, step up one by one until we've found a valid prefix, then keep stepping from where we left off.

The problem is in the line

```
encoded = encoded[length:]
```

Slicing a string like this creates a copy of the string. And creating a copy of string takes a number of operations that scales like the length of the string.

This means for each starting point (the number of which is roughly linear in the length of the string) we're making a copy of a large chunk of the string (which is itself roughly linear in the length of the string), which means that the whole operation is quadratic. Let's see how it scales with input size:

```
N = 10
bigstr = ''.join(encoder[math.gcd(15, n)] for n in range(1, N))
%timeit c = chunk1(bigstr)
```

I get

```
N = 100              0.0573 ms
N = 1000             0.436  ms (8x)
N = 10000            6.53   ms (15x)
N = 100000         274      ms (41x)
N = 1000000      35100      ms (128x)
```

To avoid this "accidentally quadratic" behavior, we need to avoid making the expensive copy. Instead of lopping off the front of the string, we can just leave the string untouched and use a start variable to keep track of where to begin:

```
def chunk2(encoded: str) -> List[str]:
    chunks: List[str] = []

    start = 0
    while start < len(encoded):
        # Try length 1, length 2, ...
        for end in itertools.count(start + 1):
            prefix = encoded[start:end]

            if prefix in decoder:
                chunks.append(prefix)
                start = end
                break  # out of for loop

    return chunks
```

Let's check that it passes the same tests:

```
chunks = chunk2(f"{bignum:0>167b}")
#                         as-is   as-is   fizz   as-is   buzz
assert chunks[:5] == [ '0',    '0',   '10',   '0',   '110']
```

And now let's check the timings:

```
10              0.0039 ms
100             0.0368 ms   (9x)
1000            0.402  ms   (11x)
10000           4.25   ms   (11x)
100000         43.3    ms   (10x)
1000000       402      ms   (9x)
```

Much better! Multiply the input size by 10, and the function runs about 10 times as long.

This seems like such a minor change, but when you're chunking an encoding of the first million Fizz Buzz results it's the difference between it taking half a second and it taking 35 seconds.

Of course, this is not my preferred solution either. I'd prefer to use regular expressions.

Regular Expressions

Regular expressions provide a way of finding text that matches certain conditions. For example, the regular expression '[a-z]' matches a single lowercase character, whereas the regular expression '^(dog|cat)$' matches the string "dog" or the string "cat" (the ^ is an anchor meaning "start of string" and the $ is an anchor meaning "end of string").

The functions for working with regular expressions are in the re library, which we imported as part of our example. The three most common (at least, the three I use the most) are re.search, re.match, and re.findall.

re.search tells you if any part of the string matches the regular expression:

```
assert re.search('[a-z]', 'a')
assert re.search('[a-z]', 'baBY')
assert not re.search('[a-z]', '01234')
```

re.match tells you if the *beginning* of the string matches the regular expression:

```
assert re.match('[a-z]', 'a')
assert re.match('[a-z]', 'baBY')
assert not re.match('[a-z]', 'BaBY')
assert not re.match('[a-z]', '01234')
```

Both of those are not just true/false checks but (if they match) return re.Match objects that contain information about the details of the match.

Our example used re.findall, which returns all non-overlapping matches:

```
assert re.findall('[a-z]', 'baBy') == ['b', 'a', 'y']
assert re.findall('[a-z]', '01234') == []
assert re.findall('aa', 'aaaaa') == ['aa', 'aa']
```

Notice that in the last example there are only two non-overlapping instances of the substring aa. This suggests the regex-based solution to our problem – we define a regex that matches any of our codes:

```
huffman_regex = '(0|10|110|111)'
```

Regular expressions match in a *greedy* way, so it will attempt to match the start of the string, if possible. Due to the nature of our code, exactly one of those codes will match the start of our string. And due to the non-overlapping nature of re.findall, the matching will resume right after whatever just matched, which is exactly the behavior we want:

```
chunks = re.findall('(0|10|110|111)', f"{bignum:0>167b}")

assert len(chunks) == 100
assert chunks[:5] == ['0', '0', '10', '0', '110']
```

As a mild bonus, this is about 3x faster than our `chunks2` implementation, although it scales in the same linear way.

The only thing left is to finish the decoding. Instead of using `decoder` (which would map back to the gcd, which we'd then have to so something with), we can just interpret the result as a binary number. (Luckily for us, all the codes represent different binary numbers.)

```
'0'   -> binary 0 -> as-is
'10'  -> binary 2 -> fizz
'110' -> binary 6 -> buzz
'111' -> binary 7 -> fizzbuzz
```

One solution would be to use a dict:

```
[{0: str(n+1), 2: 'fizz', 6: 'buzz', 7: 'fizzbuzz'}[int(c, 2)]
  for n, c in enumerate(chunks)]
```

But I chose a more oblique solution, which was to use a list with garbage values at the unused indices:

```
[[str(n+1), '', 'fizz', '', '', '', 'buzz', 'fizzbuzz'][int(c, 2)]
  for n, c in enumerate(chunks)]
```

7. Infinite Iterables

One day I was thinking about `itertools`, as I am prone to do, and this elegant solution struck me out of the blue:

```python
import itertools

fizzes = itertools.cycle(['', '', 'fizz'])
buzzes = itertools.cycle(['', '', '', '', 'buzz'])
numbers = itertools.count(1)

fizz_buzzes = ((fizz + buzz) or str(n)
               for fizz, buzz, n in zip(fizzes, buzzes, numbers))

output = [next(fizz_buzzes) for _ in range(100)]
```

In this chapter we'll explore how this solution works.

Iterables, Iterators, and Lazy Infinite Sequences

Typically, one of the first things you learn in Python is lists:

```python
xs = ['a', 'b', 'c', 'd', 'e']
```

There are a couple of common ways to access the elements of a list. The first is to retrieve an element at a specific position:

```
# lists are indexed starting at 0
assert xs[2] == 'c'
```

The second is to *iterate* over the list, for example using a `for` loop:

```
for x in xs:
    assert 'a' <= x <= 'e'
```

But lists are not the only things you can iterate over. For example, we've used `enumerate` to get the elements of a list along with their indices. But it's certainly not a list:

```
es = enumerate(xs)
```

```
try:
    es[0]
except TypeError:
    print("'enumerate' object is not subscriptable")
```

In fact, there is a much broader class of objects ("iterables") that can be iterated over. It turns out that anything can be iterated over if it knows how to generate an "iterator".

The distinction in Python between iterators and iterables is subtle and confusing (in particular, because every iterator is an iterable but not vice versa) and so we'll go through it in somewhat excruciating detail.

Iterators

Something is an *iterator* if you can call `next()` on it. Calling `next()` will either return some "next" element, or it will raise a `StopIteration` exception. You can think of an iterator as a stream of elements that can be traversed only once, in order, and that may end at some point.

Why only once in order? Because the only way to access the elements of an iterator is by calling `next()` to get the next element. There is no way to "go back" nor to access any element that's not the "next" one.

One iterator that you may have dealt with in Python is a generator expression:

```
# generator containing 1, 2, 3
it = (x for x in [1, 2, 3])

assert next(it) == 1  # next element is 1
assert next(it) == 2  # next element is 2
assert next(it) == 3  # next element is 3
try:
    next(it)
    assert False
except StopIteration:
    print("no more elements")
```

Another is a function that yields values:

```
from typing import Iterator

def upto(n: int) -> Iterator[int]:
    for i in range(n):
        yield i

it = upto(3)

assert next(it) == 0
assert next(it) == 1
assert next(it) == 2

try:
    next(it)
    assert False
except StopIteration:
    pass
```

In practice you rarely call next() yourself (although we will sometimes be doing so in this chapter), and so in practice you rarely work with iterators as iterators.

Iterables

Something is an *iterable* if it knows how to give you an iterator. In particular, every iterator is an iterable, since it can just give you itself.

You call the `iter()` function on an iterable to get its iterator. When you call it on an iterator it just returns itself:

```
# it is an iterator and therefore also an iterable
it = (x for x in [1, 2, 3])

# iterator is its own iterator
assert iter(it) == it
```

Of course you can also get iterators for other iterables:

```
# xs is not an iterator, calling next(xs) would give an error
xs = [1, 2, 3]

# but it is an iterable
it = iter(xs)

assert next(it) == 1
assert next(it) == 2
assert next(it) == 3
```

Notice that while the iterator is consumed, the list is not. At any time you can get a fresh iterator, or even multiple iterators at once:

```
xs = [1, 2, 3]
it1 = iter(xs)
it2 = iter(xs)
assert next(it1) == next(it2) == 1
assert next(it1) == next(it2) == 2
```

Typically we'll work using tools like for loops. Behind the scenes you could imagine them doing something like:

```python
from typing import Iterable, Callable

def for_each(xs: Iterable[int], do: Callable) -> None:
    it = iter(xs)

    try:
        while True:
            do(next(it))
    except StopIteration:
        pass

for_each([1, 2, 3], print)  # works with lists
for_each(upto(3), print)    # works with iterators
```

Working with the iterable abstraction allows us not to care whether we have lists or generators or whatever.

Infinite Iterables

So far all of the iterables we've used eventually stop. But it's easy to make an iterator that goes on forever:

```python
def every_number() -> Iterator[int]:
    n = 0
    while True:
        yield n
        n += 1
```

You shouldn't use a for loop with this iterator, since `for x in every_number()` would never stop (unless your loop had a `break` in it, or something similar).

Instead you have to use other techniques. For example, you could define a function to take the first n results from an iterable and return them in a list:

```python
from typing import List

def head(xs: Iterable[int], n: int) -> List[int]:
    results = []
    it = iter(xs)

    for _ in range(n):
        try:
            results.append(next(it))
        except StopIteration:
            break

    return results
```

This works on infinite iterables:

```python
assert head(every_number(), 5) == [0, 1, 2, 3, 4]
```

and finite ones:

```
assert head([2, 3, 4, 5], 2) == [2, 3]
```

Working with iterables in this way is somewhat clumsy. Luckily, there's a better way.

itertools

Python comes with the `itertools` library, a collection of functions for working with iterators and iterables. Generally speaking, these functions will take iterables (or nothing) as inputs and return iterators as outputs.

For example, itertools contains an `islice` function which is roughly the equivalent of the `head` function we wrote in the previous section (except that it returns an iterator not a list):

```
import itertools

def head2(xs: Iterable[int], n: int) -> List[int]:
    return list(itertools.islice(xs, n))

assert head(every_number(), 5) == head2(every_number(), 5)
assert head([2, 3, 4], 2) == head2([2, 3, 4], 2)
```

The functions in `itertools` do many interesting things, and you are encouraged to read the documentation and/or check out my 2015 presentation on Stupid Itertools Tricks for Data Science.

Here we only need two of its functions.

The first is `itertools.count` which returns an infinite iterator that "counts" based on the supplied `start` (default: 0) and `step` (default: 1). That is, `itertools.count()` is the iterator that generates the values 0, 1, 2, ... and keeps going.

```
assert head(itertools.count(), 5) == [0, 1, 2, 3, 4]
assert head(itertools.count(start=10, step=2), 3) == [10, 12, 14]
```

The other one we'll need is `itertools.cycle`, which takes an iterable and returns its elements repeatedly in a cycle forever:

```
guesses = itertools.cycle(['yes', 'no', 'maybe'])
assert head(guesses, 5) == ['yes', 'no', 'maybe', 'yes', 'no']
```

`zip` VS `map`

`zip` is a commonly used function in Python. It lazily "zips" together multiple iterables into an iterator of tuples, stopping whenever the first is exhausted. (There is an `itertools.zip_longest` in case you want it to keep going as long as one of the iterables is not exhausted.)

So, for example,

```
pairs = zip([1, 2], ['a', 'b', 'c'])
assert list(pairs) == [(1, 'a'), (2, 'b')]

triplets = zip([1, 2], ['a', 'b', 'c'], [True, False])
assert list(triplets) == [(1, 'a', True), (2, 'b', False)]
```

One way to combine multiple iterables is using `zip` and a list comprehension. For example,

```python
names = ['Alice', 'Bob', 'Carol']
titles = ['Professor', 'Doctor', 'President']

greetings = [f"{title} {name}"
             for title, name in zip(names, titles)]

assert greetings = [
    "Professor Alice",
    "Doctor Bob",
    "President Carol"
]
```

Another way is to use `map`, which takes a mapping function and some iterables to use as its arguments one value at a time:

```python
greetings2 = map(lambda title, name: f"{title} {name}",
                 names,
                 titles)

assert list(greetings2) == greetings
```

I generally avoid `map`, for a couple of reasons. One is that it returns a lazy iterable, so that you have to remember to wrap it in `list()` if you actually want a list. (`zip` has the same issue, but you typically use it inside a `for` loop, so it doesn't end up mattering.)

The second reason is that I find Python code that uses `map` to be "unpythonic"; that is, non-idiomatic. This is a judgment call on my part, but I'm the one writing the book.

Putting It All Together

How do we use all these concepts to solve the problem? First, we define

```
# An infinite iterator whose every third element is 'fizz'
# and the rest of whose elements are the empty string
fizzes = itertools.cycle(['', '', 'fizz'])
```

and

```
# An infinite iterator whose every fifth element is 'buzz'
# and the rest of whose elements are the empty string
buzzes = itertools.cycle(['', '', '', '', 'buzz'])
```

and finally

```
# The infinite iterator 1, 2, 3, 4, ...
numbers = itertools.count(1)
```

Now consider the iterator zip(fizzes, buzzes, numbers). Its first several elements are

```
('', '', 1)
('', '', 2)
('fizz', '', 3)
('', '', 4)
('', 'buzz', 5)
('fizz', '', 6)
```

and its 15th element is ('fizz', 'buzz', 15).

That is, whenever the number is divisible by 3, the first element will be 'fizz', and whenever the number is divisible by 5 the second element will be 'buzz', which means that whenever the number is divisible by 15 the first two elements will be 'fizz' and 'buzz'.

If the first two elements are not both empty, the correct answer is their concatenation (since 'fizz' + '' == 'fizz'). If they are both empty, the right answer is str() of the number. We can get this with

```
fizz_buzzes = ((fizz + buzz) or str(n)
                for fizz, buzz, n in zip(fizzes, buzzes, numbers))
```

This works because or returns its left-hand value if that value is truthy, and its right-hand value otherwise. Here, if either `fizz` or `buzz` is non-empty, `fizz + buzz` is a non-empty string (which is truthy) and so that's what gets returned. If both `fizz` and `buzz` are empty strings, `fizz + buzz` is an empty string (which is not truthy), and so `str(n)` gets returned.

Notice that since the inputs are infinite iterables we have to use a generator comprehension. If we used a list comprehension it would try to materialize this infinite sequence as a list and would run forever.

This actually solves more than the original problem, since it generates an infinite stream of Fizz Buzz outputs. We only need the first 100:

```
output = [next(fizz_buzzes) for _ in range(100)]
```

Bonus: PowerFizz

Here's a variant I call "PowerFizz":

> Print the numbers 1 to N, except that if the number is a perfect square, print "fizz"; if the number is a perfect cube, print "buzz"; and if the number is a sixth power, print "fizzbuzz".

The obvious solution doesn't work because of the usual floating point reasons:

```python
def is_sixth_power(n: int) -> bool:
    sixth_root = n ** (1 / 6)
    return sixth_root == int(sixth_root)

assert is_sixth_power(1 ** 6)
assert is_sixth_power(2 ** 6)
assert is_sixth_power(3 ** 6)
assert not is_sixth_power(4 ** 6)  # !!
```

Of course, you could use the try-every-number method:

```python
def is_kth_power(n: int, k: int) -> bool:
    for i in itertools.count(1):
        if i ** k == n:
            return True
        # Once we pass n, it's not a k-th power.
        elif i ** k > n:
            return False

assert is_kth_power(10 ** 6, 6)
assert is_kth_power(3 ** 5, 5)
assert not is_kth_power(3 ** 4, 5)
```

But we can also easily solve this variant using iterators. We create a squares iterator consisting of all squares and a cubes iterator consisting of all cubes. Any time n equals the first element of squares we know n must be a square, and then we advance that iterator and are ready to find the next square. Any time n equals the first element of cubes we know n must be a cube, and then we advance that iterator and are ready to find the next cube.

There's no easy way to "peek" at the first element of an iterator, so we just use next() to get the first element and store it in a "buffer" variable:

```python
def power_fizz() -> Iterator[str]:
    squares = (n ** 2 for n in itertools.count(1))
    cubes = (n ** 3 for n in itertools.count(1))

    # Buffer for next element of each iterator
    next_square, next_cube = next(squares), next(cubes)

    for n in itertools.count(1):
        fizz = buzz = ''
        if n == next_square:
            fizz = 'fizz'
            next_square = next(squares)
        if n == next_cube:
            buzz = 'buzz'
            next_cube = next(cubes)
        yield (fizz + buzz) or str(n)
```

When n is a square, `fizz` is `"fizz"`; otherwise it's an empty string. When n is a cube, `buzz` is `"buzz"`; otherwise it's an empty string. And when n is a sixth power (that is, both a square and a cube) then `fizz` is `"fizz"` and buzz is `"buzz"`, as desired:

```python
# Put a None at the beginning so output[1] is the output for 1
output = [None] + head(power_fizz(), 1000)

assert output[7 ** 2] == "fizz"
assert output[6 ** 3] == "buzz"
assert output[2 ** 6] == "fizzbuzz"
assert output[2 ** 7] == str(2 ** 7)
```

8. Random Guessing

Python has a `random` module for generating pseudo-random numbers. Its `choice` function randomly picks an element from a list. For example – `random.choice([0, 1, 2, 3])` returns 0, 1, 2, or 3, all with equal probability.

Surprisingly, we can use `random.choice` to solve FizzBuzz:[1]

```python
from typing import Iterator
import random
import itertools

def fizz_buzzes() -> Iterator[str]:
    counts = [itertools.count(1)] * 15
    for group in zip(*counts):
        random.seed(23_977_775)
        for n in group:
            # Just pick at random
            yield random.choice(
                ['fizzbuzz', 'fizz', str(n), 'buzz']
            )

fb = fizz_buzzes()
output = [next(fb) for _ in range(100)]
```

How could this possibly work?

random

Computers are good at following instructions exactly. They are not great at being random. This means that libraries like Python's `random` are not truly generating

---

[1]I learned about this method from a Stack Overflow post.

random numbers; rather, they are generating numbers that "seem" random, or are random enough for most purposes.

For example, if you want to simulate rolling dice a lot of times, `random` is probably random enough. You can simulate a single die roll by using `random.randint()`:

```python
# Don't worry about this, we'll explain it later
random.seed(12)

def roll(num_dice: int) -> int:
    """'rolls' `num_dice` 6-sided 'dice' and returns the total"""
    return sum(random.randint(1, 6)
               for _ in range(num_dice))
```

Let's create a histogram of a lot of rolls:

```python
from collections import Counter

counts = Counter(roll(2) for _ in range(10_000))

# histograms on the cheap:
for value, count in sorted(counts.items()):
    print(f"{value:2} {count:5} {'*' * (count // 100)}")
```

We get a distribution that looks like

```
 2    306 ***
 3    537 *****
 4    869 ********
 5   1089 **********
 6   1368 *************
 7   1623 *****************
 8   1416 **************
 9   1098 **********
10    840 ********
11    565 *****
12    289 **
```

Other functions you might use are `random.random`, which gives you a number drawn uniformly from [0, 1); `random.randrange`, which gives you a number drawn uniformly from the corresponding `range`, `random.choice`, which picks a random element from a list; and `random.shuffle`, which randomly shuffles a list (in place).

As you saw, the Fizz Buzz solution in this chapter used `random.choice`.

Pseudorandomness and the Seed

The `random` module doesn't actually generate random numbers; it generates *pseudorandom* numbers. That is, it starts with some "seed" and generates a deterministic series of numbers.

Nonetheless, every time you run a Python program that uses `random`, you will probably get different results. For example, if from the command line you repeatedly run

```
python -c"import random; print(random.randrange(100))"
```

you will get a random-looking series of numbers, not the same number every time.

This is because by default Python sets the seed based on your computer's time. The second time you run this the time will be different, which means the seed will be different, and so the sequence of "random" numbers will be different.

However, if you specify the same seed, the numbers will be the same:

```
python -c"import random; random.seed(12);
print(random.randrange(100))"
```

will print "60" every time you run it (at least until Python changes its random number generation algorithm).

If you're rolling "dice" for your tabletop fantasy game, you don't need to set the seed yourself; however, if you're trying to make your program run in a *reproducible* way you definitely should. The examples in the next few sections will set the seed just so that you can get the same results as I do. (And the solution in this chapter sets the seed for reasons we'll talk about later.)

Creating a Pseudorandom Generator

It's not hard to create our own pseudo-random number generator. We'll create what's called a linear congruential generator that starts with some value x (the seed) and then keeps computing x = (a * x + c) % m for some carefully chosen constants a, c, and m:

```
import time

# A "global variable" to represent the random seed.
class Random:
    seed = int(time.time())


def random_number() -> int:
    m = 2 ** 32
    a = 1664525
    c = 1013904223

    Random.seed = (a * Random.seed + c) % m
    return Random.seed
```

In general it is considered poor form to use and modify global variables in Python. One alternative would be for seed to be an *instance-level* variable, and that we be required to create an *instance* of Random in order to generate random numbers. In some ways this would be a cleaner approach. (It would also allow us to maintain multiple separate stateful random number generators.)

However, this would require us to explicitly reference and pass around the random state, which would be clunky and muddy up our exposition. So we'll bite our tongues and use the single global Random.seed. At least it's encapsulated as a class-level variable.

So now let's think about the random_number function. It generates a single number between 0 and m $-$ 1, that is, 2 $**$ 32 $-$ 1. That's usually not exactly what we want. For instance, we might prefer to generate a number uniformly between 0 and some n (not including n).

One way to do that is to split the interval [0, 2 ** 32) into n buckets and take the index of the bucket each random number falls into. The problem is that for most n we won't be able to find equally sized buckets. For example, 2 ** 32 is not divisible into three equal parts. And if our buckets are not equally sized, our random numbers are not uniform, since they're more likely to fall into a larger bucket than a smaller one.

We deal with this by making the buckets as large as we can (while keeping them equal-sized) and then throwing away random numbers that fall outside the last bucket:

```python
def random_range(n: int) -> int:
    # Find the largest bucket size that fits
    bucket_size = 2 ** 32 // n

    while True:
        draw = random_number()
        idx = draw // bucket_size
        # If the number falls outside the last bucket, try again.
        if idx < n:
            return idx
```

And then we can repeat our dice rolling example:

```
Random.seed = 12

counts = Counter()

for _ in range(10_000):
    # random_range is between 0 and n-1, so we need to add 1
    die1 = random_range(6) + 1
    die2 = random_range(6) + 1
    counts[die1 + die2] += 1

for value, count in sorted(counts.items()):
    print(f"{value:2} {count:5} {'*' * (count // 100)}")
```

This results in a reasonable-looking histogram:

```
 2    271 **
 3    596 *****
 4    860 ********
 5   1124 ***********
 6   1390 *************
 7   1636 ****************
 8   1387 *************
 9   1070 **********
10    823 ********
11    558 *****
12    285 **
```

Similarly, we can use this to generate uniform randoms between 0 and 1. We'll count how many fall in each decile:

```python
def uniform_random() -> float:
    return random_number() / 2 ** 32

Random.seed = 12

counts = Counter()

for _ in range(10_000):
    # Truncate down to multiple of 0.1
    decile = int(uniform_random() * 10) / 10
    counts[decile] += 1

for value, count in sorted(counts.items()):
    print(f"{value:2} {count:5} {'*' * (count // 100)}")
```

which gives roughly uniform-looking deciles:

```
0.0   1015 **********
0.1   1103 ***********
0.2   1005 **********
0.3    948 *********
0.4    965 *********
0.5    988 *********
0.6    996 *********
0.7   1001 **********
0.8    977 *********
0.9   1002 **********
```

This seems reasonable. Nonetheless, there are more sophisticated tests for randomness that this will fail, and this is not the algorithm by which Python generates random numbers.

Advances in the Same Iterator

As part of our solution, we'll need to "chunk" an iterable into groups of fixed size. So it's worth thinking about how to accomplish this.

Let's start with the specific case of splitting a string into chunks of length 2. Here's a possibly surprising solution:

```python
from typing import List

def chunks_of_2(s: str) -> List[str]:
    it = iter(s)
    return [''.join(pair) for pair in zip(it, it)]

assert chunks_of_2("0123456789") == ['01', '23', '45', '67', '89']
```

It might not be obvious how this works. Let's unpack it by writing our own toy version of `zip`:

```python
def toyzip(xs, ys):
    # Get an iterator for each of xs and ys
    xit = iter(xs)
    yit = iter(ys)

    try:
        while True:
            x = next(xit)
            y = next(yit)
            yield x, y
    except StopIteration:
        return
```

Think about what happens if we pass in the *same* iterator for both arguments:

```
l = [1, 2, 3, 4, 5, 6]
it = iter(l)

assert list(toyzip(it, it)) == [(1, 2), (3, 4), (5, 6)]
```

Within the function, xs and ys both equal the iterator it. Calling iter() on an iterator just gives you back the iterator itself, which means that xit and yit also both equal the iterator it. They are the *same object*. Calling next(xit) is the same as calling next(it). Calling next(yit) is the same as calling next(it).

That is, the first call to next(xit) returns 1 and advances the iterator. The first call to next(yit) (which is the same iterator, which was just advanced) returns 2 and advances the iterator. Those lines are taking turns advancing the same iterator, which means that next(xit) gets 1, 3, and 5; and next(yit) gets 2, 4, and 6. Hence this trick gives us the values pairwise.

What if we wanted values in groups of 5? In groups of arbitrary size? Read on.

Values and References

One common class of bug in Python programs results from accidentally using multiple instances of the same object instead of copies of that object:

```
# Make a row of 3 zeros
row = [0, 0, 0]

# Use three copies to make a 3x3 matrix
matrix = [row] * 3

# Set the top left element to 1
matrix[0][0] = 1

print(matrix)
# [[1, 0, 0], [1, 0, 0], [1, 0, 0]]
```

This is probably not what you expected to happen!

The problem is that [row] * 3 does not actually make three copies; it's the equivalent of writing [row, row, row]. Each row is the exact same object. That is, matrix[0], matrix[1], and matrix[2] are all references to the same list object. Changing the first element of one necessarily changes the first element of all of them, because they're all the same thing.

Although this often causes bugs, we can sometimes use this to our advantage:

```
# This is 5 references to the same iterator
counts = [itertools.count(1)] * 5

assert [next(count) for count in counts] == [1, 2, 3, 4, 5]
assert [next(count) for count in counts] == [6, 7, 8, 9, 10]
```

If we didn't want this behavior, we could use a list comprehension:

```
# This is 5 different iterators
counts = [itertools.count(1) for _ in range(5)]

assert [next(count) for count in counts] == [1, 1, 1, 1, 1]
assert [next(count) for count in counts] == [2, 2, 2, 2, 2]
```

Argument Unpacking

Functions like zip can take arbitrarily many arguments. Most commonly we give zip two arguments:

```
for x, y in zip([1, 2, 3], [4, 5, 6]):
    print(x + y)
```

But it's happy to accept three or four or more:

```
s = "0123456"

four_grams = [
    ''.join(four_tuple)
    for four_tuple in zip(s, s[1:], s[2:], s[3:])
]

assert four_grams == ["0123", "1234", "2345", "3456"]
```

What if we wanted to give it an arbitrary number of arguments? This won't do the right thing:

```
strings = ["cat", "dog", "emu"]

triplets = list(zip(strings))

# :( wanted to get ('c', 'd', 'e'), ('a', 'o', 'm'), ('t', 'g', 'u')
assert triplets == [("cat",), ("dog",), ("emu",)]
```

Here zip took the list of strings and performed a "single-input zip"; that is, it returned the elements of that list as 1-tuples.

To treat the elements as separate arguments, we can use a * to "unpack" the list:

```
trip2 = list(zip(*strings))

assert trip2 == [('c', 'd', 'e'), ('a', 'o', 'm'), ('t', 'g', 'u')]
```

The * tells Python to take the elements out of the list and use them as individual function arguments.

We can similarly write a function that expects any number of arguments:

```
def toyzip2(*xses):
    its = [iter(xs) for xs in xses]

    while True:
        try:
            values = [next(it) for it in its]
        except StopIteration:
            return

        yield tuple(values)

tuples = list(toyzip2([1, 2], [3, 4, 5], [6, 7]))
assert tuples == [(1, 3, 6), (2, 4, 7)]
```

And indeed our solution uses zip in this fashion.

The Cycle of 15, Redux

The solution in this chapter takes advantage of the fact that random.choice is actually deterministic. This means that (in theory) if we choose the right seed we can produce any desired sequence of random numbers.

Now, there are 4 ** 100 possible sequences of 100 random choices from [0, 1, 2, 3]. The first element can be any of 0, 1, 2, or 3; as can the second element; as can the third element, and so on.

```
print(4 ** 100)
# 1606938044258990275541962092341162602522202993782792835301376
```

That's an *extremely* large number. The odds of finding a seed that will produce the exact sequence we want are miniscule.

However, as we saw in the cycle of 15, we really only need to solve the problem for the numbers 1 through 15, after which the same solution repeats forever.

That is, if we can find a seed that gets us 15 correct random draws, we can keep resetting the seed and get the same 15 correct random draws again and again.

Where did 23_977_775 come from?

There are 4 ** 15 possible outcomes of calling

```
[random.randrange(4) for _ in range(15)]
```

This is approximately a billion combinations:

```
# 1,073,741,824
print(f"{4 ** 15:,}")
```

This means that if you wanted to produce a specific sequence, and if you tried a billion different seeds, you'd expect that approximately one of them would give rise to the "random" sequence you wanted.

Here we want a random sequence that corresponds to the "cycle of 15" representing the four Fizz Buzz cases. Throughout the book, we've mostly been treating "as-is" as case 0, "fizz" as case 1, "buzz" as case 2, and "fizzbuzz" as case 3. In which case the sequence we're hoping for would be

```
target = [0, 0, 1, 0, 2, 1, 0, 0, 1, 2, 0, 1, 0, 0, 3]
```

However, there's no reason why we have to label the cases this way. We could just have easily have made "fizzbuzz" case 0 and "as-is" case 3. There are 4! = 24 different ways to label the cases. One way to see this is to notice that we have 4 ways to label "fizzbuzz", then for each such choice we have 3 ways to label "buzz", and so on: 4 * 3 * 2 * 1 = 24.

We'd be happy with a seed that produces any of these 24 sequences, which we should expect to happen if we check 4 ** 15 / 24, or approximately 44 million seeds. This is a much more manageable number. How can we generate those sequences?

```
import itertools

remaps = [dict(enumerate(permutation))
          for permutation in itertools.permutations([0, 1, 2, 3])]
```

Here itertools.permutations generates each of the 24 permutations. One of these is (1, 3, 0, 2). enumerate turns this into an iterable of pairs (0, 1), (1, 3), (2, 0), (3, 2) which dict then turns into a "remap":

```
{0: 1, 1: 3, 2: 0, 3: 2}
```

We can then use each of these to remap our desired sequence:

```
targets = [[remap[i] for i in target] for remap in remaps]
assert len(targets) == 24
```

Here [remap[i] for i in target] just replaces the values in target based on the specified remap. In the example above, each 0 would be replaced by 1, each 1 by 3, and so on, resulting in

```
[1, 1, 3, 1, 0, 3, 1, 1, 3, 0, 1, 3, 1, 1, 2]
```

which corresponds to the order ["buzz", "as-is", "fizzbuzz", "fizz"].

Now we simply need to find a random seed that generates any one of our targets:

```
seed = 0
generated = None
while generated not in targets:
    seed += 1
    random.seed(seed)
    generated = [random.choice([0, 1, 2, 3]) for _ in range(15)]
print(seed, generated)

# 23977775 [2, 2, 1, 2, 3, 1, 2, 2, 1, 3, 2, 1, 2, 2, 0]
```

This tells us we need to use index 0 for "fizzbuzz", index 1 for "fizz", index 2 for "as-is", and index 3 for "buzz", which is exactly what we did.

Putting it all together

There's quite a few things going on here.

First, we make 15 copies of the same iterator:

```
counts = [itertools.count(1)] * 15
```

Then we zip them together to form groups of 15:

```
for group in zip(*counts):
```

At the start of each group of 15 we reset the random seed:

```
random.seed(23_977_775)
```

And then we make 15 random draws whose results we already know:

```
for n in group:
    yield random.choice(
        ['fizzbuzz', 'fizz', str(n), 'buzz']
    )
```

Simple!

9. Matrix Multiplication

I hope you like linear algebra:

```python
import numpy as np

def fizz_buzz(n: int) -> str:
    w = np.array([[1, 0, 0], [2, -2, 0], [2, 0, -2], [3, -3, -3]])
    v = np.array([1, n % 3, n % 5])
    return [str(n), 'fizz', 'buzz', 'fizzbuzz'][np.argmax(w @ v)]
```

In this chapter we'll think about Fizz Buzz as a matrix multiplication problem.

n-Dimensional Arrays

An *n-dimensional array* is a collection of numeric elements with a specific structure. The dimension n indicates how many indexes you need to specify to indicate which array element you want.

A 0-dimensional array is essentially just a number. You don't need any indexes to specify which value you want because there's only a single value.

```python
zero_d_array = 10
```

A 1-dimensional array is a fixed-size list of numbers. You need only one index to specify a single value – its position in the list. (Sometimes we call a 1-dimensional array a *vector*, and sometimes we just refer to it as an *array*.)

```
vector = [-1, 0, 1]
another_vector = [0.9, 0.5, 0.4, 0.6]

assert vector[2] == 1, "element at index 2 is 1"
```

A 2-dimensional array is (in some sense) a list of equally-sized 1-dimensional arrays. We often refer to a two-dimensional array as a *matrix*:

```
identity_matrix = [
    [1, 0, 0],
    [0, 1, 0],
    [0, 0, 1]
]

non_square_matrix = [
    [1, 2, 3, 4],
    [5, 6, 7, 8]
]

assert non_square_matrix[1][1] == 6
```

We can keep going: a 3-dimensional array is a stack of matrices, and so on:

```
# shape (3, 3, 3)
three_d_tic_tac_toe = [
    [[-1, 0, 0], [1, 0, 0], [0, 0, 0]],
    [[0, -1, 0], [0, 0, 1], [0, 0, 0]],
    [[-1, 0, 0], [0, 0, 0], [0, 1, 0]],
]
```

The *shape* of an n-dimensional array is the n-tuple consisting of the sizes of its dimensions, outermost-to-innermost. So vector has shape (3,), non_square_matrix has shape (2, 4), and three_d_tic_tac_toe has shape (3, 3, 3).

Digression: Recursion, Iteration, and Tail Recursion

How would you compute the shape of an arbitrary n-dimensional array where you don't necessarily know what n is? The problem has a nice recursive structure to it:

- find the size of the outermost dimension
- prepend it to the shape of one of the rows

For example, in our `non_square_matrix`, you'd first notice that it has two rows (the outermost dimension). Then you'd look at the shape of one of those rows. Each row is an array of size 4, and then there are no dimensions remaining.

Sounds simple enough, but how do we know when there's no dimensions remaining? Well, the way we've represented things, once we reach something that's no longer a list, we're out of dimensions.

```python
from typing import Tuple, Union

def shape(ndarray: Union[list, float]) -> Tuple[int, ...]:
    if isinstance(ndarray, list):
        # More dimensions, so make a recursive call
        outermost_size = len(ndarray)
        row_shape = shape(ndarray[0])
        return (outermost_size, *row_shape)
    else:
        # No more dimensions, so we're done
        return ()
```

As always, we write a few tests to make sure it works:

```
assert shape(zero_d_array) == ()
assert shape(vector) == (3,)
assert shape(non_square_matrix) == (2, 4)
assert shape(three_d_tic_tac_toe) == (3, 3, 3)
```

Recursion isn't always the best way to solve problems in Python. Each time we make a recursive call to shape Python has to remember where it was called from (and what all the local variables were equal to) so that it can resume execution when that recursive call returns.

Python uses a *call stack* to keep track of these recursive calls, but it's only allowed to reach a certain size. If the call stack gets too big, Python will raise a RecursionError:

```
def blow_the_stack(n: int = 0) -> None:
    try:
        blow_the_stack(n + 1)
    except RecursionError:
        print(f"crashed after {n} calls")

blow_the_stack()
```

For me this crashes after about 2900 recursive calls.

Now, you'll probably never be working with 2900-dimensional arrays, and so the recursive version of shape should be fine. (If each array dimension had size 2, such an array would have 2 ** 2900 elements, which is a 873-digit number.)

Nonetheless, it's easy enough to change our shape function to eliminate the use of recursion. One way to do this is by managing the "call stack" ourselves. Here the call stack is really only used to remember each size to prepend to the results. So instead we can just stick those sizes in a list as we go:

```
def shape2(ndarray: Union[list, float]) -> Tuple[int, ...]:
    sizes = []

    while isinstance(ndarray, list):
        # append outermost size to `sizes`
        sizes.append(len(ndarray))
        # handle "recursion" by replacing ndarray with ndarray[0]
        ndarray = ndarray[0]

    # instead of prepending we can just return the sizes
    return tuple(sizes)
```

In our original recursive implementation, after the recursive call returned we had to add its result to another tuple and return that. That is, even after the recursive call returned, there's was more work to do with its result.

An alternative way to write such functions is using *tail recursion*, a style in which the function makes a recursive call but just returns the result directly.

Typically this involves adding an additional "accumulator" argument to the function:

```
def shape3(ndarray: Union[list, float],
           sizes: Tuple[int, ...] = ()) -> Tuple[int, ...]:
    if isinstance(ndarray, list):
        outermost_size = len(ndarray)
        return shape3(ndarray[0], sizes + (outermost_size,))
    else:
        return sizes
```

Rather than hanging on to outermost_size and prepending it once the recursive call returns, we instead pass it (and all the sizes preceding it) into the function.

Why is this difference interesting? Now the recursive call is in the "tail position"; that is, we are just returning its result directly. This means there's no longer any need to remember where it was called from. We don't need to remember the value of outermost_size, and we don't need to prepend anything to the result of the recursive call. We just need to return the result of the recursive call.

Some languages can do what's called *tail-call optimization.* That means that behind the scenes they rewrite tail-recursive functions to run in a loop, which means you can recurse as deep as you want.

Alas, the standard implementation of Python doesn't do this, so it's no help to us. In any event, we don't actually need to write our own shape function, because we're not going to be working with lists-as-arrays.

NumPy

In Python to work with n-dimensional arrays you most frequently use the numpy library:

```
python -m pip install numpy
```

It's a standard convention to alias the import as np:

```
import numpy as np
```

In fact, it's so standard that people will think you're strange if you don't alias the import this way.

You can create arrays with the np.array function.

```
zero_d_array = np.array(10)

vector = np.array([-1, 0, 1])

non_square_matrix = np.array([[1, 2, 3, 4], [5, 6, 7, 8]])
```

Why do we need a special library? Why can't we just work with lists?

One reason is that numpy stores and uses arrays efficiently. It knows what type of objects an array contains and is able to create a smart representation behind the scenes and use optimized C code for its operations. This enables numpy to do mathematics much faster. For example, imagine having 1 million random numbers and wanting to sum them:

```
import random

list_1m = [random.random() for _ in range(1_000_000)]
array_1m = np.array(list_1m)
```

Summing the list is (relatively) slow:

```
# about 5 milliseconds
%timeit sum(list_1m)
```

Using numpy's array.sum is more than 10 times faster:

```
# about 300 microseconds
%timeit array_1m.sum()
```

For more complex operations the differences are even more stark.

Another reason is that there are lots of things you'd like to do with arrays, and numpy has them implemented in an efficient way already. For example, there's no need to write our own shape function:

```
assert non_square_matrix.shape == (2, 4)
```

Working With Arrays

There are lots and lots of things you can do with arrays and with numpy. In this section we'll talk about only a few.

dtypes

As mentioned earlier, each numpy array has a specific data type, which is called its dtype:

```
assert np.array([1, 2, 3]).dtype == np.int64
```

Here, numpy has inferred that the array is an array of 64-bit integers. If one of them were a floating-point number, it would have inferred that type:

```
assert np.array([1.0, 2, 3]).dtype == np.float64
```

You can also specify the dtype explicitly if numpy would infer the wrong one:

```
assert np.array([1, 2, 3], dtype=np.float64).dtype == np.float64
```

Once the type is set, it's set:

```
int_array = np.array([1, 2, 3])        # dtype = np.int64
int_array[0] = 1.1                     # assign a float?
assert int_array[0] == 1               # nope, it got truncated
assert int_array.dtype == np.int64     # dtype is the same
```

Arithmetic

If two arrays have the same dimensions, you can do arithmetic on them elementwise.

```
a1 = np.array([1, 2, 3, 4])
a2 = np.array([5, 6, 7, 8])

np.testing.assert_array_equal(a1 + a2, np.array([6, 8, 10, 12]))
np.testing.assert_array_equal(a1 * a2, np.array([5, 12, 21, 32]))
```

Note that I didn't use Python's `assert`. That's because `a1 == a2` is the array of elementwise comparisons (here, an array of four `False` values) and numpy won't let you treat multi-element arrays as booleans. That is, if you tried to do something like `assert a1 + a2 == a3`, numpy would raise an exception even if the arrays were "equal".

Instead I used one of numpy's `testing.assert` functions.

In some circumstances numpy can "broadcast" a smaller array against a larger one; for example:

```python
small = np.array([-1, 0, 1])
large = np.array([[10, 20, 30], [40, 50, 60]])

np.testing.assert_equal(small + large,
                        np.array([[9, 20, 31], [39, 50, 61]]))
```

In this case (because the dimensions work out nicely) `small` gets "broadcast" and added to each row of `large`. However, if `large` were a 2x2 matrix, this broadcasting wouldn't be possible, and you'd get an error.

Dot Products

One common operation that's not just element-wise arithmetic is computing the *dot product* of two same-sized vectors, which is just the sum of the products of corresponding elements:

```python
from typing import List

def dot_product(v1: List[float], v2: List[float]) -> float:
    assert len(v1) == len(v2)
    return sum(x1 * x2 for x1, x2 in zip(v1, v2))
```

But of course we'll use `np.dot`, which is orders of magnitude faster.

Why is the dot product interesting? It has various geometric interpretations, but for us the most useful way to think about it is as a way of computing weighted sums.

Imagine that we have a vector of length 3: v = [x0, x1, x2]. Then the dot product with [1, 0, 0] is just x0. It's large when x0 is large and small when x0 is small.

Similarly, the dot product with [0, 1, 0] is x1, and the dot product with [0, 0, 1] is x2.

What about the dot product with [1, -1, -1]? This equals x1 - x2 - x3, and it's large when x1 is much bigger than the sum of x2 and x3, and small when the opposite is true.

We'll use this idea in our solution.

Matrix Multiplication

If M is a matrix with n rows and m columns, and if v is a vector of size m, then we can define their product as the vector of size n each of whose elements is the dot product of v with the corresponding row of M:

```
def matrix_times_vector(m, v):
    return [dot_product(row, v) for row in m]
```

And if N is another matrix, with m rows and k columns, we can define the matrix product of M and N as the (n, k) matrix whose element at (i, j) is the dot product of the i-th row of M with the j-th column of N. (Both of those have m elements, so the dot product makes sense.) We won't bother implementing it ourselves.

In Python we use the @ operator for matrix multiplication. Matrix multiplication is not actually implemented in base Python, but the operator calls the special __matmul__ method of an object, and numpy has thoughtfully implemented this method for us to use.

This means we can do the following:

```
m1 = np.array([
    [10, 20, 30, 40],
    [14, 13, 12, 11]
])
a1 = np.array([1, 2, 3, 4])
product = m1 @ a1
```

which behaves as we described:

```
# (2, 4) matrix times (4,) array => (2,) array
assert product.shape == (2,)

# product[0] should be the dot product of a1 with m[0]
assert product[0] == (1 * 10) + (2 * 20) + (3 * 30) + (4 * 40)

# product[1] should be the dot product of a1 with m[1]
assert product[1] == (1 * 14) + (2 * 13) + (3 * 12) + (4 * 11)
```

Fizz Buzz as a Matrix Multiplication Problem

In our solution we defined a (3,) vector that depends on n:

```
v = np.array([1, n % 3, n % 5])
```

and multiplied it by a constant (4, 3) matrix:

```
w = np.array([1, 0, 0], [2, -2, 0], [2, 0, -2], [3, -3, -3])
```

The product is a vector of size 4, corresponding to our Fizz Buzz classes.

We then used `np.argmax` to find the index of the largest element and took that as our class. But why does this work? Why does the product of that matrix and that vector produce the correct largest element?

Let's go through the rows of the weights matrix one at time.

The first row is `[1, 0, 0]`. The dot product of this row with v is always 1, since the first element of v is always 1. This means the output corresponding to "as-is" is always 1.

The second row is `[2, -2, 0]`. The dot product of this row with v is

```
2 - 2 * (n % 3)
```

If n is divisible by 3, n % 3 is 0 and the dot product of this row with v equals 2, which is larger than the "as-is" output. Otherwise, n % 3 is 1 or 2, so the dot product equals 0 or -2. In particular, when n is not divisible by 3, the dot product is smaller than the "as-is" output.

The third row is `[2, 0, -2]`. Similar reasoning shows that this row's dot product equals 2 if n is divisible by 5, and that it's 0 or smaller if n is not divisible by 5.

The last row is `[3, -3, -3]`. Its dot product with v is

```
3 - 3 * (n % 3) - 3 * (n % 5)
```

If both n % 3 and n % 5 are 0, then this equals 3. Otherwise it's 0 or less.

What does this mean for the four Fizz Buzz cases?

If n is divisible by neither 3 nor 5, then the first element is 1 and the other three elements are all 0 or smaller. So in the as-is case the first element is largest.

If n is divisible by 3 but not by 5, then the second element is 2 and the other three elements are less than 2. So in the "fizz" case the second element is largest.

If n is divisible by 5 but not by 3, then similar reasoning shows that the third element is largest.

And if n is divisible by both 3 and 5, then the last element is 3, and the other elements are always less than 3. So in the "fizzbuzz" case the last element is largest.

That means this produces the outcome we want.

argmax

The final element of our solution was np.argmax, which gives the index of the largest element. If the largest element is at index 0, it returns 0. If the largest element is at index 3, it returns 3. That's exactly the behavior we need:

```
def fizz_buzz(n: int) -> str:
    w = np.array([1, 0, 0], [2, -2, 0], [2, 0, -2], [3, -3, -3])
    v = np.array([1, n % 3, n % 5])

    # 0 if as-is, 1 if fizz, 2 if buzz, 3 if fizzbuzz
    idx = np.argmax(w @ v)
    return [str(n), 'fizz', 'buzz', 'fizzbuzz'][idx]
```

And that's our solution.

How Did I Choose Those Weights?

Well, mostly by trial and error. But here was roughly my thought process.

The input vector is [1, n % 3, n % 5]. The first element will always be 1. The second will be 0 if n is divisible by 3, and 1 or 2 otherwise. The third will be 0 if n is divisible by 5, and between 1 and 4 otherwise.

I didn't really want to have to worry about the different values of n % 3 and n % 5, only whether each was 0 or not. So I set out to create weights that only cared about that.

I thought it would be simplest if I could have the as-is output be constant, so I set it to 1 and tried to get the other outputs to be more than or less than 1 as appropriate.

In order to beat that in the "fizz" case I wanted the "fizz" output to be 2 when n was divisible by 3 and 0 or less otherwise. This was easy enough, I just started with 2 and subtracted off 2 times n % 3.

I did basically the same thing for the "buzz" row. I wasn't worried about a tie, because if both the "fizz" and "buzz" outputs were 2, then n was divisible by both 3 and 5, and I would need "fizzbuzz" to be larger than those anyway.

That left just the "fizzbuzz" case where both n % 3 and n % 5 were zero. To beat the "fizz" and "buzz" outputs I wanted the "fizzbuzz" output to be 3. And subtracting off 3 times n % 3 and n % 5 ensures the output is zero or less if either n % 3 or n % 5 is not zero.

10. Fizz Buzz in Tensorflow

In this chapter we'll look at Fizz Buzz as a *machine learning* problem. Unlike the other chapters, it doesn't lend itself to a simple solution that I can show off and then explain, so you'll have to bear with me for a bit.

We'll be using the `pytorch` library for working with neural networks. This makes the title of this chapter misleading, but it is what it is for historical reasons. (One such reason is that PyTorch did not yet exist when I wrote the original "Fizz Buzz in Tensorflow" blog post.)

You can install it with

```
python -m pip install torch>=1.4.0
```

Machine Learning

In each of our solutions so far, we've constructed the *logic* for how to solve Fizz Buzz and instructed the computer to follow our instructions.

Machine learning is an alternative approach to solving problems where instead of writing out a solution, we get the computer to *learn* a solution from data.

For example, imagine the problem of identifying whether an image is a picture of a cat or a dog. Coming up with explicit logic to do this would be pretty hard. But collecting lots of labeled pictures of cats and dogs is relatively easy; if the computer could just somehow learn to distinguish them our job would be done.

In this chapter, rather than reasoning out how to solve Fizz Buzz (which you should be able to do blindfolded at this point), we'll give the computer "solved" examples and try to get it to learn the underlying pattern.

Fizz Buzz as a Machine Learning Problem

In Matrix Multiplication our solution involved multiplying a vector of features by a matrix of weights. In that chapter I reasoned out what the weights should be in order to get the correct result. So that wasn't "machine learning". But what if instead we had learned the weights from data? Let's try that using PyTorch.

At a high level, here's how this kind of learning works:

1. We take a batch of inputs and their corresponding correct labels.
2. We apply a parametrized *model* to the inputs to get a prediction.
3. We apply a *loss function* to the predictions and the correct labels to measure how "good" the predictions are.
4. PyTorch knows how the loss function depends on the model weights, and it uses an *optimizer* to adjust the weights in a way that makes the loss smaller.

If we repeat this process many times, if we have representative training data, and if we've chosen a reasonable model for the problem, then hopefully it should learn to predict pretty well.

We'll start with the matrix multiplication model we used previously. Matrix multiplication is a pretty fundamental building block of machine learning. Accordingly, PyTorch comes with a `torch.nn.Linear` module that has a matrix of weights that it multiplies with its input. (By default it also adds a constant, so we'll need to set `bias=False`.)

We're treating Fizz Buzz as a *multiclass classification* problem: there are four output classes, and for each input there is a single correct class.

In the last chapter we used `argmax` to turn the output vector into a prediction. But for this kind of machine learning problem we want a loss function (and hence a prediction) that depends in a smooth way on the weights and inputs. For this reason it's common to interpret the (softmax of the) outputs as a probability distribution over the classes, and to use `torch.nn.CrossEntropyLoss` as the loss function to minimize.

Finally, we'll try to find the optimal matrix of weights using `torch.optim.SGD`, the stochastic gradient descent optimizer.

Here we already know that matrix multiplication can solve the problem, so we expect to find weights that work, although they are likely to be different from our previous bespoke weights.

Let's try it. First we'll define the model, loss function, and optimizer:

```
import torch

torch.manual_seed(12)  # for reproducibility

# Set bias=False so it's just a matrix multiplication
model = torch.nn.Linear(in_features=3, out_features=4, bias=False)
loss = torch.nn.CrossEntropyLoss()
optimizer = torch.optim.SGD(model.parameters(), lr=0.1)
```

For each number n we'll need to produce a `torch.Tensor` of features and the correct class:

```
import math

def make_features(n: int) -> torch.Tensor:
    # Use 1.0 to force a tensor of floats
    return torch.tensor([1.0, n % 3, n % 5])

def fizz_buzz_class(n: int) -> int:
    return [1, 3, 5, 15].index(math.gcd(n, 15))
```

We'll be using the `torch.utils.data.DataLoader` to batch and shuffle our data. Luckily, it works well with `NamedTuples`, so we can create a `NamedTuple` class to represent the instances in our dataset.

Each instance will represent a number n, and will contain a tensor of `features` as well as the `label` corresponding to the correct Fizz Buzz class (here 0, 1, 2, or 3).

We'll also give it a static "factory" method that automatically computes the features and the label, so that we can do

```
Instance.create(15, make_features)
```

instead of having to do

```
Instance(15, make_features(15), fizz_buzz_class(15))
```

This design makes it easy to experiment with different ways of generating features.

```
from typing import NamedTuple, Callable

class Instance(NamedTuple):
    n: int
    features: torch.Tensor
    label: int

    @staticmethod
    def create(n: int, featurize: Callable) -> 'Instance':
        return Instance(n, featurize(n), fizz_buzz_class(n))
```

We can now use this to generate training data:

```
training_data = [Instance.create(n, make_features)
                 for n in range(1, 101)]
```

And then we'll train 500 passes over the data:

```
from torch.utils.data import DataLoader

# Each epoch is one pass through the data
for epoch in range(500):
    epoch_loss = 0.0

    batches = DataLoader(training_data,
                         batch_size=5,
                         shuffle=True)
    for batch in batches:
        # Reset the optimizer before each step
        optimizer.zero_grad()

        predictions = model(batch.features)

        # Compute the error
        error = loss(predictions, batch.label)
        # Backpropagate the errors
        error.backward()
        # Add this loss to the total epoch loss
        epoch_loss += error.item()
        # Tweak the weights to reduce the error
        optimizer.step()
    print(epoch, epoch_loss)

# print out the model weights
print(model.weight)
```

When I run this, the loss steadily decreases and I get weights

```
Parameter containing:
tensor([[-3.7892,   4.2138,   4.1363],
        [ 0.2882,  -6.4787,   4.5541],
        [ 0.0583,   4.7219,  -5.2487],
        [ 3.9191,  -3.5162,  -3.4348]], requires_grad=True)
```

You can run through the cases as before to convince yourself that these weights are good ones.

Let's also write a function to evaluate how well a model predicts on a dataset:

```
def evaluate(model: torch.nn.Module,
             data: list,
             verbose = False) -> int:
    num_correct = 0

    # Don't compute gradients when evaluating
    with torch.no_grad():
        for n, features, label in data:
            predicted = torch.argmax(model(features)).item()
            num_correct += predicted == label

            if verbose:
                check = "√" if predicted == label else "×"
                outputs = [str(n), 'fizz', 'buzz', 'fizzbuzz']
                print(check, n, outputs[predicted], outputs[label])

    return num_correct
```

If you evaluate the model we just trained:

```
evaluate(model, training_data, verbose=True)
```

you'll find that it gets 100/100 correct.

Learning and Generalization

When we solve a problem with machine learning, we're typically using specific training examples to learn a solution to a more general problem.

If you have 100 pictures of cats and 100 pictures of dogs, it is pretty easy to "learn" a function that can distinguish cats from dogs on those 100 pictures. (For example, you could memorize which filenames correspond to "dog" and which to "cat".)

It's a much harder problem to learn to tell cats from dogs *even in pictures that weren't part of the training set*. If you can do that well across a wide range of new-to-you pictures, then it's more likely you're actually good at solving the problem.

In this sense, the matrix weights we learned in the previous section were "cheating". We used the training examples from 1 to 100 to learn a model to predict the correct outputs for the numbers 1 to 100.

In this particular case it doesn't really matter, since the `features` function already contains most of the solution (we'll discuss this shortly), but in real life you should not use your training data to evaluate how good your model is.

You can verify that it doesn't matter by changing the training data to be the numbers 101 to 1000:

```
training_data = [Instance.create(n, make_features)
                 for n in range(101, 1000)]
```

You'll find that you get a similar-looking weights matrix and that it still gets 100/100 correct on the test set 1 to 100.

Feature Engineering

The other way in which we "cheated" was that our `make_features` function already contained most of the solution. The canonical solution from if / elif / elif / else mostly involved computing n % 3 == 0 and n % 5 == 0 and then taking a simple action based on the results.

The matrix multiplication model we just trained did much the same. But we pretty much gave it "divisible by 3" and "divisible by 5" as its *inputs*, and all it had to learn was the "simple action" corresponding to each combination.

Another way to think about this is that the features we chose would only work for this exact problem. Imagine instead that "fizz" was for multiples of 2 and "buzz" for multiples of 7:

```
def fizz_buzz_27_class(n: int) -> int:
    return [1, 2, 7, 14].index(math.gcd(n, 14))
```

If you create instances with these labels you'll find that the same model is unable to learn much of anything at all. This is because the features aren't a good representation of the inputs; they mostly just represent the (3, 5, 15) outputs.

Imagine instead that we didn't *know* exactly the details of the problem we needed to solve; we just knew that we needed to train a model to classify numbers somehow. And that we wanted enough flexibility that the model could learn (with different weights, of course) various different classification schemes. In this case using

```
[1, n % 3, n % 5]
```

as the features wouldn't make much sense at all.

So, what can we use as a more general way to turn a number into a fixed-size array of features, suitable for handing off to a matrix multiplication model?

One-Hot Decimal Digits

One easy way is to *one-hot encode* the decimal digits of each number. That is, represent each digit as an array of length 10 that is mostly zeros except for a single 1 in the position corresponding to the digit.

We'll start with a function that can encode a single digit:

```python
from typing import List

def one_hot_digit(digit: int) -> List[float]:
    """
    Return a list of length 10, that's all zeros
    except for result[digit] == 1
    """
    assert 0 <= digit <= 9, "digit must be between 0 and 9"

    return [1.0 if i == digit else 0.0 for i in range(10)]
```

As always we'll check that it works:

```python
assert one_hot_digit(0) == [1, 0, 0, 0, 0, 0, 0, 0, 0, 0]
assert one_hot_digit(7) == [0, 0, 0, 0, 0, 0, 0, 1, 0, 0]
```

And then we'll use that function to encode multiple-digit numbers:

```python
def one_hot_digits(n: int, max_digits: int = 3) -> torch.Tensor:
    output = []
    for _ in range(max_digits):
        digit = n % 10
        output.extend(one_hot_digit(digit))
        n = n // 10

    # The digits are in reverse order, but it doesn't matter
    return torch.tensor(output)
```

```
assert one_hot_digits(72).tolist() == [
    0, 0, 1, 0, 0, 0, 0, 0, 0, 0,   # 2
    0, 0, 0, 0, 0, 0, 0, 1, 0, 0,   # 7
    1, 0, 0, 0, 0, 0, 0, 0, 0, 0,   # 0
]
```

That is, we turn each input into an array of length 30 that contains exactly 3 ones representing its last 3 (decimal) digits.

We need to change the `in_features` of the model to accommodate this.

```
torch.manual_seed(12)   # for reproducibility

model = torch.nn.Linear(in_features=30, out_features=4)
optimizer = torch.optim.SGD(model.parameters(), lr=0.01)
```

And we should create new datasets with these features:

```
one_hot_training_data = [Instance.create(n, one_hot_digits)
                         for n in range(101, 1000)]

one_hot_test_data = [Instance.create(n, one_hot_digits)
                     for n in range(1, 101)]
```

If you train this model, you'll find that it doesn't learn very much. For me, it gets 67/100 correct. That sounds pretty good, until you realize that the rule "if the last digit is 0 or 5 say 'buzz', otherwise say the number as-is" also gets 67/100 correct. (My trained model is not doing exactly this, but it's doing something similar.)

And such a rule is extremely easy to learn from this representation, since it boils down to "if `features[0]` or `features[5]` is 1, say 'buzz'; otherwise say 'as-is'."

Why can't this representation learn more than that? One problem is that it's very sparse. Although each input is 30-dimensional, each only has a non-zero value in three places. This results in a lot of "wasted space" that the model can't do anything with.

Another problem is that it's not capturing much about the *structure* of the numbers. For instance, imagine (a) changing the last digit of a number from 3 to 6, and (b) changing the last digit of a number from 4 to 7.

Both of these result in adding three to the number, but from the perspective of this representation they have absolutely nothing to do with one another: one involves turning `inputs[3]` off and `inputs[6]` on, the other involves turning `inputs[4]` off and `inputs[7]` on. This means that anything a model learns about the first (which would be reflected in the weights it applies to `inputs[3]` and `inputs[6]`) it's unlikely to be able to apply to the second.

Binary Digits

What's a denser way to represent each number? As we learned in Decimal, Binary, and Hexadecimal, we can use the binary representation:

```python
def binary_digits(n: int, num_digits: int = 10) -> torch.Tensor:
    digits = []
    for _ in range(num_digits):
        # Need to use floats
        digits.append(float(n % 2))
        n = n // 2
    return torch.tensor(digits)
```

Let's check that we did it right:

```python
assert (binary_digits(0).tolist() ==
        [0, 0, 0, 0, 0, 0, 0, 0, 0, 0])
assert (binary_digits(10).tolist() ==
        [0, 1, 0, 1, 0, 0, 0, 0, 0, 0])
assert (binary_digits(1020).tolist() ==
        [0, 0, 1, 1, 1, 1, 1, 1, 1, 1])
```

Now we can represent the same set of numbers (actually, all numbers up to 1023) using only a 10-dimensional input.

```
binary_training_data = [Instance.create(n, binary_digits)
                        for n in range(101, 1023)]

binary_test_data = [Instance.create(n, binary_digits)
                    for n in range(1, 101)]
```

If I train the matrix model with with these features (after changing the `in_features` to 10 and cranking the learning rate down to 0.001), and using training examples from 101 to 1000, I get a model that always predicts as-is and so gets 53/100 correct.

Deep Learning

Maybe the problem is the model?

Our simple matrix multiplication model can only compute linear combinations of the inputs. We can compute richer outputs by stacking together linear layers with non-linear "activation" layers into a *neural network*:

```
input_dim = 10        # for binary encoding
hidden_dim = 5        # we'll vary this
torch.manual_seed(12) # so you get the same results

model = torch.nn.Sequential(
    # Linear layer: input_dim -> hidden_dim
    torch.nn.Linear(in_features=input_dim, out_features=hidden_dim),
    # ReLU(x) = max(x, 0)
    torch.nn.ReLU(),
    # Linear layer: hidden_dim -> 4
    torch.nn.Linear(in_features=hidden_dim, out_features=4)
)
```

We'll use the AdamW optimizer:

```
optimizer = torch.optim.AdamW(model.parameters())
```

and train for 2500 epochs:

```
for epoch in range(2500):
    epoch_loss = 0.0

    for batch in DataLoader(binary_training_data,
                            batch_size=5,
                            shuffle=True):
        optimizer.zero_grad()

        predictions = model(batch.features)
        error = loss(predictions, batch.label)
        error.backward()
        epoch_loss += error.item()
        optimizer.step()

    num_correct = evaluate(model,
                           binary_test_data,
                           verbose=epoch % 100 == 0)
    print(f"epoch: {epoch:>5} "
          f"accuracy: {num_correct}/100 "
          f"loss: {epoch_loss:.2f}")

evaluate(model, binary_test_data, verbose=True)
```

How does it do? Well, "always pick as-is" gets 53/100 correct. Hopefully any model can learn that rule, so we'll use that as our baseline.

With 5 hidden dimensions, I can get up to about 58/100 correct. This is better than always guessing "as-is", but not by much. Still, it means that the model is learning *something*.

With 10 hidden dimensions I can get it to learn around 65/100 correct. If I go up to 25 hidden dimensions, I can get to about 81/100 correct. At 50 dimensions, I can get to about 88/100 correct. And you could keep going, create more complicated architectures, add dropout, and so on. If your goal was to get to 100/100, I bet you could get there.

Validation

Although we're using a training dataset that's distinct from the test dataset, we're still using *performance on the test dataset* both to track the progress of our learning and to choose among different models.

It turns out this is a different form of cheating. Stopping training when we get no more improvement on the test dataset and choosing a model based on performance on the test dataset means that we're still (somewhat implicitly) using the test data to "train" our model.

A more hygienic approach would be to split our data into separate train / validate / test datasets, train the model using the training data, evaluate the training processes using the validation data, and then evaluate our final chosen model once on the test data.

Here our goal is simply to explore whether and how a machine learning model can solve the Fizz Buzz problem, so I'm ignoring this kind of discipline, but I feel somewhat guilty about doing so.

How Does It Work?

The 25-hidden-dimension model nails all of the "fizzbuzz" cases: it says "fizzbuzz" for 15, 30, 45, 60, 75, and 90; and never says it otherwise. So it's worth looking at how it does that.

To start with, let's generate the predictions for all numbers 1 to 1023. As usual we'll create a NamedTuple so that we can easily pair each instance with its prediction:

```
class Prediction(NamedTuple):
    instance: Instance
    output: torch.Tensor
```

Then we can generate all the predictions:

```
# here `model` is the trained 25-dimensional model

all_data = binary_test_data + binary_training_data

with torch.no_grad():
    predictions = [Prediction(instance, model(instance.features))
                   for instance in all_data]
```

Let's start by looking at the instances with the highest values for the "fizzbuzz" output:

```
# Sort by highest "fizzbuzz" logit
predictions.sort(key=lambda p: p.output[3], reverse=True)

def show(prediction: Prediction) -> None:
    print(prediction.instance.n,
          prediction.instance.features.int().tolist(),
          [round(x, 1) for x in prediction.output.tolist()])

for prediction in predictions[:10]:
    show(prediction)
```

For me this produces

```
# n            input features                    outputs
#    1  2  4  8 16 32 64 128 256 512 as-is fizz  buzz fizzbuzz
450 [0, 1, 0, 0, 0, 0, 1, 1,  1,  0] [3.4, -2.2, -5.8,  9.0]
210 [0, 1, 0, 0, 1, 0, 1, 1,  0,  0] [3.5, -2.1, -6.0,  8.9]
195 [1, 1, 0, 0, 0, 0, 1, 1,  0,  0] [3.6, -2.1, -6.1,  8.5]
975 [1, 1, 1, 1, 0, 0, 1, 1,  1,  1] [3.7, -2.3, -5.8,  8.4]
990 [0, 1, 1, 1, 1, 0, 1, 1,  1,  1] [3.7, -2.3, -5.8,  8.2]
735 [1, 1, 1, 1, 1, 0, 1, 1,  0,  1] [3.6, -2.2, -5.7,  8.2]
960 [0, 0, 0, 0, 0, 0, 1, 1,  1,  1] [3.7, -1.9, -6.5,  7.7]
720 [0, 0, 0, 0, 1, 0, 1, 1,  0,  1] [3.8, -1.9, -6.7,  7.7]
495 [1, 1, 1, 1, 0, 1, 1, 1,  1,  0] [3.7, -2.2, -5.7,  7.5]
255 [1, 1, 1, 1, 1, 1, 1, 1,  0,  0] [3.7, -2.1, -5.7,  7.4]
```

The two largest "fizzbuzz" outputs are for 450 and 210. In fact, their entire output vectors are almost identical. Their binary representations are also similar: they only differ in two bits. 450 has the 256 bit on and the 16 bit off. And 210 has the 256 bit off and the 16 bit on.

In other words,

```
450 = 210 + 256 - 16
    = 210 + 240
```

195 also has very similar outputs. And its binary representation also only differs in two bits from 210:

```
210 = 195 + 16 - 1
    = 195 + 15
```

The next three (975, 990, 735) have similar outputs to each other and follow the same pattern:

```
990 = 975 + 16 - 1
    = 975 + 15

975 = 735 + 256 - 16
    = 735 + 240
```

And the next two, 960 and 720, also follow the +256 -16 pattern in their binary representations, as do 495 and 255.

Notice also that all these one-bit-on-one-bit-off differences are multiples of 15. (Although here that isn't surprising, since all these inputs are multiples of 15, which means that their differences necessarily will be too.)

Let's next look at the highest "buzz" scores:

```
# Sort by highest "buzz" logit
predictions.sort(key=lambda p: p.output[2], reverse=True)

for prediction in predictions[:10]:
    show(prediction)
```

which gives

```
20  [0, 0, 1, 0, 1, 0, 0, 0, 0, 0] [-2.3,  0.1, 4.0, -11.9]
5   [1, 0, 1, 0, 0, 0, 0, 0, 0, 0] [-2.4,  0.1, 4.0, -11.6]
260 [0, 0, 1, 0, 0, 0, 0, 0, 1, 0] [-2.3,  0.0, 3.9, -11.9]
895 [1, 1, 1, 1, 1, 1, 1, 0, 1, 1] [-1.6,  0.7, 3.7,  -8.1]
470 [0, 1, 1, 0, 1, 0, 1, 1, 1, 0] [-1.6, -0.0, 2.9, -12.5]
35  [1, 1, 0, 0, 0, 1, 0, 0, 0, 0] [-1.0, -0.0, 2.9,  -6.0]
455 [1, 1, 1, 0, 0, 0, 1, 1, 1, 0] [-1.7, -0.0, 2.8, -11.6]
215 [1, 1, 1, 0, 1, 0, 1, 1, 0, 0] [-1.7,  0.0, 2.8, -11.5]
115 [1, 1, 0, 0, 1, 1, 1, 0, 0, 0] [-1.3, -0.3, 2.7,  -6.6]
355 [1, 1, 0, 0, 0, 1, 1, 0, 1, 0] [-1.5, -0.1, 2.6,  -6.8]
```

Again we see a similar pattern: the top three numbers (all of which have extremely similar outputs) differ by a +16 -1 bit flip and a +256 -16 bit flip. And there are two more +256 -16 pairs: 455 / 215 and 355 / 115.

Let's continue with "fizz":

```
# Sort by highest "fizz" logit
predictions.sort(key=lambda pair: pair[1][1], reverse=True)

for prediction in predictions[:10]:
    show(prediction)
```

Here we see a lot more variation:

```
513 [1, 0, 0, 0, 0, 0, 0, 0, 0, 1] [-0.3, 1.3, -0.1, -8.2]
378 [0, 1, 0, 1, 1, 1, 1, 0, 1, 0] [-0.2, 1.3, -1.3, -5.9]
786 [0, 1, 0, 0, 1, 0, 0, 0, 1, 1] [ 0.2, 1.2, -0.7, -6.6]
603 [1, 1, 0, 1, 1, 0, 1, 0, 0, 1] [ 0.0, 1.2, -0.8, -7.1]
519 [1, 1, 1, 0, 0, 0, 0, 0, 0, 1] [ 0.7, 1.2, -2.6, -7.4]
18  [0, 1, 0, 0, 1, 0, 0, 0, 0, 0] [-0.2, 1.1, -0.3, -6.4]
534 [0, 1, 1, 0, 1, 0, 0, 0, 0, 1] [ 0.8, 1.1, -2.7, -7.5]
843 [1, 1, 0, 1, 0, 0, 1, 0, 1, 1] [ 0.1, 1.1, -0.7, -7.0]
258 [0, 1, 0, 0, 0, 0, 0, 0, 1, 0] [-0.2, 1.1, -0.4, -6.1]
723 [1, 1, 0, 0, 1, 0, 1, 1, 0, 1] [ 0.2, 1.1, -0.7, -7.8]
```

You can see we have

```
843 = 603 + 256 - 16
    = 603 + 240
```

but in general we're not quite seeing the same patterns as above. Possibly we've reached the point where looking at "highest output logit" simply doesn't give us similar outputs. What if we looked for similar outputs directly?

That is, given some number n we want to find the other numbers for which the model produces the "most similar" outputs. Here we'll measure similarity by "distance in 4-dimensional space", which we can compute with np.linalg.norm.

To make our lives easier, we'll write a distance_from function that takes as input a prediction p and returns another function that gives the distance from p to its input:

```python
import numpy as np

def distance_from(p: Prediction):
    # Create function to return
    def distance_from_p(p2: Prediction):
        return np.linalg.norm(p.output - p2.output)

    return distance_from_p
```

Let's start by creating a lookup table so we can quickly find the prediction for a given n:

```python
lookups = {prediction.instance.n: prediction
           for prediction in predictions}
```

Then for any n we can find the most similar outputs by sorting all the predictions by their distance from the prediction for n.

```python
def most_similar_outputs(n: int):
    predictions.sort(key=distance_from(lookups[n]))

    for prediction in predictions[:10]:
        show(prediction)
```

The results for 513 are not very interesting, but if we do most_similar_outputs(378), we see

```
378 [0, 1, 0, 1, 1, 1, 1, 0, 1, 0] [-0.2,  1.3, -1.3, -5.9]
558 [0, 1, 1, 1, 0, 1, 0, 0, 0, 1] [ 0.3,  0.9, -1.2, -5.7]
123 [1, 1, 0, 1, 1, 1, 1, 0, 0, 0] [-0.1,  0.9, -0.8, -5.4]
498 [0, 1, 0, 0, 1, 1, 1, 1, 1, 0] [-0.2,  1.1, -0.6, -6.2]
306 [0, 1, 0, 0, 1, 1, 0, 0, 1, 0] [-0.4,  1.1, -0.8, -5.3]
363 [1, 1, 0, 1, 0, 1, 1, 0, 1, 0] [-0.0,  0.9, -0.8, -5.3]
858 [0, 1, 0, 1, 1, 0, 1, 0, 1, 1] [ 0.0,  1.0, -0.5, -6.0]
258 [0, 1, 0, 0, 0, 0, 0, 0, 1, 0] [-0.2,  1.1, -0.4, -6.1]
894 [0, 1, 1, 1, 1, 1, 1, 0, 1, 1] [ 0.6,  0.9, -1.1, -5.8]
39  [1, 1, 1, 0, 0, 1, 0, 0, 0, 0] [ 0.2,  0.5, -1.5, -6.2]
```

which you can see contains

```
123 = 378 - 256 + 1  = 378 - 255
498 = 378 + 128 - 8  = 378 + 120
363 = 378 - 16  + 1  = 378 - 15
858 = 378 + 512 - 32 = 378 + 480
```

Notice that these "bit flip differences" are all still multiples of 15, which is interesting and suggestive.

If we try a random as-is input, say 91, the most similar output is:

```
91  [1, 1, 0, 1, 1, 0, 1, 0, 0, 0] [1.7, 0.0, -4.1, -0.8]
331 [1, 1, 0, 1, 0, 0, 1, 0, 1, 0] [1.6, 0.0, -4.0, -1.0]
```

Again, we have

```
331 = 91 + 256 - 16
    = 91 + 240
```

And here's 89:

```
89  [1, 0, 0, 1, 1, 0, 1, 0, 0, 0] [1.6, -0.7, -4.9, -4.0]
58  [0, 1, 0, 1, 1, 1, 0, 0, 0, 0] [1.9, -0.6, -4.7, -4.1]
329 [1, 0, 0, 1, 0, 0, 1, 0, 1, 0] [1.5, -0.7, -4.6, -4.3]
```

As we're starting to expect, we have

```
329 = 89 + 256 - 16
    = 89 + 240
```

What does this all mean?

Remember from Cycle of 15 that if two numbers differ by a multiple of 15 then they have the same Fizz Buzz output. So a model that could learn to ignore differences that were multiples of 15 would be well-positioned to learn how to Fizz Buzz.

As we mentioned before, there's no obvious way for a neural network trained on binary representations to learn "divisible by 15".

However, imagine that some part of the neural network puts the same weights on the "1" bit and on the "16" bit. That means that if we have two inputs that are identical except that one has the 1-bit on but not the 16-bit, and the other has the 16-bit on but not the 1-bit, then the corresponding parts of the network behave identically for those two inputs. 210 and 195 ("fizzbuzz") are one such pair. 20 and 5 ("buzz") another. And there are many ways to get differences that are multiples of 15 using such bit flips:

```
# pairs (i1, i2) where i1 > i2 and
# 2 ** i1 - 2 ** i2 is a multiple of 15
for i1 in range(10):
    for i2 in range(i1):
        n1, n2 = 2 ** i1, 2 ** i2
        if (n1 - n2) % 15 == 0:
            print(f"{n1} - {n2} = {n1 - n2}")
```

which gives

```
16  -  1  =  15
32  -  2  =  30
64  -  4  =  60
128 -  8  =  120
256 -  1  =  255
256 - 16  =  240
512 -  2  =  510
512 - 32  =  480
```

That is, whenever a number has only the larger of those bits set in its binary representation, we can subtract that multiple of 15 from it by swapping the two bits. Whenever it has only the smaller of those bits set, we can add that multiple of 15 by swapping the two bits.

This suggests a handwavy explanation of what the neural network might be doing:

1. "memorize" the correct outputs for some small number of inputs
2. learn to ignore "bit flips" that correspond to multiples of 15
3. profit

Let's write a function to show how our predictions change as we perform these bit flips. If we are right, they will change very little when the bit flips represent multiples of 15 but much more when they don't.

For a given input n we'll find all the other numbers between 1 and 1023 that we can get to by flipping two bits, and we'll sort them by how close their predictions are to the prediction for n:

```python
def bit_flip_predictions(n: int):
    digits = binary_digits(n)
    prediction = lookups[n]

    # predictions for numbers 1 to 1023 we can get from bit flips
    flip_predictions = [lookups[n - 2 ** i1 + 2 ** i2]
                        for i1 in range(10)
                        for i2 in range(10)
                        if digits[i1] and not digits[i2]
                        and 0 < n - 2 ** i1 + 2 ** i2 < 1024]

    # sort by distance from the prediction for n
    flip_predictions.sort(key=distance_from(prediction))

    # Now show the results
    for p in [prediction] + flip_predictions:
        show(p)
```

If we look at the `bit_flip_predictions` for 90 ("fizzbuzz"), we see that the closest outputs stay very similar as long as the bit flips represent multiples of 15:

```
90   [0, 1, 0, 1, 1, 0, 1, 0, 0, 0] [3.3, -1.5, -6.9, 7.0]
330  [0, 1, 0, 1, 0, 0, 1, 0, 1, 0] [3.2, -1.5, -6.7, 6.9]
75   [1, 1, 0, 1, 0, 0, 1, 0, 0, 0] [3.4, -1.4, -7.1, 6.4]
600  [0, 0, 0, 1, 1, 0, 1, 0, 0, 1] [3.1, -1.7, -5.7, 6.5]
120  [0, 0, 0, 1, 1, 1, 1, 0, 0, 0] [3.1, -1.4, -6.3, 5.5]
210  [0, 1, 0, 0, 1, 0, 1, 1, 0, 0] [3.5, -2.1, -6.0, 8.9]
30   [0, 1, 1, 1, 1, 0, 0, 0, 0, 0] [3.3, -1.7, -5.4, 4.8]
```

However, the outputs suddenly get much more different as soon as the bit flips don't represent a multiple of 15:

```
106 [0, 1, 0, 1, 0, 1, 1, 0, 0, 0] [2.4, -0.1, -4.8, -0.4]
586 [0, 1, 0, 1, 0, 0, 1, 0, 0, 1] [2.3, -0.5, -3.8, -0.4]
```

Likewise for 1 ("as is"):

```
1   [1, 0, 0, 0, 0, 0, 0, 0, 0, 0] [1.4, -0.0, -3.5, -0.3]
256 [0, 0, 0, 0, 0, 0, 0, 0, 1, 0] [1.6, -0.2, -3.6,  0.2]
16  [0, 0, 0, 0, 1, 0, 0, 0, 0, 0] [1.4, -0.2, -3.4,  0.3]

2   [0, 1, 0, 0, 0, 0, 0, 0, 0, 0] [1.6, -0.3, -2.6, -0.5]
32  [0, 0, 0, 0, 0, 1, 0, 0, 0, 0] [1.6, -0.1, -2.8, -1.5]
```

And for 80 ("buzz"):

```
80  [0, 0, 0, 0, 1, 0, 1, 0, 0, 0] [-0.9, -0.6,  0.2, -8.8]
320 [0, 0, 0, 0, 0, 0, 1, 0, 1, 0] [-1.0, -0.6,  0.4, -9.3]
65  [1, 0, 0, 0, 0, 0, 1, 0, 0, 0] [-0.9, -0.3, -0.1, -9.9]

528 [0, 0, 0, 0, 1, 0, 0, 0, 0, 1] [-0.2,  0.9,  0.3, -6.9]
18  [0, 1, 0, 0, 1, 0, 0, 0, 0, 0] [-0.2,  1.1, -0.3, -6.4]
```

And for 18 ("fizz"):

```
18  [0, 1, 0, 0, 1, 0, 0, 0, 0, 0] [-0.2, 1.1, -0.3, -6.4]
48  [0, 0, 0, 0, 1, 1, 0, 0, 0, 0] [-0.2, 1.0, -0.3, -6.1]
258 [0, 1, 0, 0, 0, 0, 0, 0, 1, 0] [-0.2, 1.1, -0.4, -6.1]
3   [1, 1, 0, 0, 0, 0, 0, 0, 0, 0] [-0.2, 1.0, -0.1, -6.2]
528 [0, 0, 0, 0, 1, 0, 0, 0, 0, 1] [-0.2, 0.9,  0.3, -6.9]

514 [0, 1, 0, 0, 0, 0, 0, 0, 0, 1] [ 1.5, 0.1, -1.4, -5.0]
6   [0, 1, 1, 0, 0, 0, 0, 0, 0, 0] [ 0.5, 0.1, -1.8, -4.5]
```

It seems that pretty much across the board the model has learned to "ignore" (that is, produce similar outputs on both sides of) bit flips that are multiples of 15 but not bit flips that aren't.

There is, of course, much more than this going on in the neural network, and you are encouraged to explore it further.

For me, however, this is a delightful discovery that reveals a hidden structure both of the Fizz Buzz problem and of binary representations, and I am pleased to end the book on such a note.

About the Author

Joel Grus is the author of the beloved book Data Science from Scratch, the beloved blog post Fizz Buzz in Tensorflow, and the polarizing JuypterCon presentation I Don't Like Notebooks.

He's also the co-host of the Adversarial Learning podcast.

He very occasionally blogs at joelgrus.com and spends most of the rest of his time on Twitter @joelgrus. If you'd like to stay up to date on his various thoughts and projects, sign up for his mailing list at joelgrus.substack.com.

And if you have feedback on the book, he would love to hear from you:

joelgrus@gmail.com

Acknowledgements

Thanks for reading! I hope you enjoyed it. Please recommend it to your friends and let me know if you have any feedback, positive or negative.

Although I have thought about Fizz Buzz quite a lot over the past four years, I never considered writing a book about it until @RoccoMeli suggested it on Twitter. Blame him for putting the idea in my head.

Thanks to Jeremy Kun and (especially) Tim Hopper for early reading and feedback. The book is much better thanks to their help.

Thanks as always to Ganga and Madeline for putting up with me writing another book.

The cover photo is from https://www.pexels.com/photo/bee-on-a-plant-2047420/ . Thank you to the photographer for allowing its free use.

Made in the USA
Middletown, DE
14 December 2021

55573971R00093